AGRICULTURAL GEOGRAPHY

LESLIE SYMONS
B.Sc.(Econ.), Ph.D.

Reader in Geography,
University College of Swansea

BELL & HYMAN LIMITED
London

First published 1967
Reprinted 1968, 1970, 1972
Revised edition 1978

ISBN 0 7135 0001 8 (cased)
0 7135 0169 3 (limp)

Printed in Great Britain by
T. & A. Constable Ltd, Edinburgh

AGRICULTURAL GEOGRAPHY

BELL'S ADVANCED ECONOMIC GEOGRAPHIES

General editor
PROFESSOR R. O. BUCHANAN
M.A.(N.Z.), B.Sc.(Econ.), Ph.D.(London)
Professor Emeritus, University of London

A. Systematic Studies

GEOGRAPHY AND ECONOMICS
Michael Chisholm, M.A.
AGRICULTURAL GEOGRAPHY
Leslie Symons, B.Sc.(Econ.), Ph.D.
THE FISHERIES OF EUROPE: AN ECONOMIC GEOGRAPHY
James R. Coull, M.A., Ph.D.
A GEOGRAPHY OF TRADE AND DEVELOPMENT IN MALAYA
P. P. Courtenay, B.A., Ph.D.
R. O. BUCHANAN AND ECONOMIC GEOGRAPHY
(Ed.) M. J. Wise, M.C., B.A., Ph.D. & E. M. Rawstron, M.A.
THE WORLD TRADE SYSTEM
SOME ENQUIRIES INTO ITS SPATIAL STRUCTURE
R. J. Johnston, M.A., Ph.D.
LAND REFORM: A WORLD SURVEY
Russell King, B.A., M.Sc., Ph.D.
AGRICULTURE IN THE THIRD WORLD: A SPATIAL ANALYSIS
W. B. Morgan, M.A., Ph.D.

B. Regional Studies

AN ECONOMIC GEOGRAPHY OF EAST AFRICA
A. M. O'Connor, B.A., Ph.D.
AN ECONOMIC GEOGRAPHY OF WEST AFRICA
H. P. White, M.A. & M. B. Gleave, M.A.
AN ECONOMIC GEOGRAPHY OF ROMANIA
David Turnock, M.A., Ph.D.
YUGOSLAVIA: PATTERNS OF ECONOMIC ACTIVITY
F. E. Ian Hamilton, B.Sc.(Econ.), Ph.D.
RUSSIAN AGRICULTURE: A GEOGRAPHIC SURVEY
Leslie Symons, B.Sc.(Econ.), Ph.D.
RUSSIAN TRANSPORT
(Ed.) Leslie Symons, B.Sc.(Econ.), Ph.D. & Colin White, B.A.(Cantab.)
AN HISTORICAL INTRODUCTION TO THE ECONOMIC GEOGRAPHY
OF GREAT BRITAIN
Wilfred Smith, M.A.
A GEOGRAPHY OF BRAZILIAN DEVELOPMENT
Janet D. Henshall, M.A., M.Sc., Ph.D. & R. P. Momsen Jr, A.B., M.A., Ph.D.

Contents

Tables

Maps and Diagrams

TO
My Wife,
Alison and Jennifer

Preface to Second Edition

When in the early 1960s I was invited to write a book on agricultural geography, I felt that my prime concern was to outline the principles and methods by which the subject should be approached and also to illustrate these by selecting particular regions for moderately detailed coverage, while not attempting a global survey of types of agriculture such as may be found in standard texts on economic geography. Since the first edition of *Agricultural Geography*, several other books with the same title have been published, and the various approaches to the same field of study have led to a reappraisal of the objectives of my own.

Basically, it remains my intention to provide a guide to what I consider to be the factors that should be taken into account in any study of agricultural distributions, and to provide enough of the substance of this knowledge to enable the book to form the basis of a course of instruction without the necessity of having at hand the resources of a large library or even the use of a large number of other texts. Hence, in Part I, after a brief historical summary of agricultural development, I included, and retain, a chapter on aspects of the physical environment that are of vital concern to farmers, and a brief survey of the main economic constraints which must be considered. References are given, however, to other texts which are examples of those needed to provide both more elementary and more advanced material. As one of my concerns is to encourage the study of agricultural geography in the context of a hungry, and potentially hungrier, world, I then, in Part II, devote four chapters to the contrasting but representative ways in which agricultural production is undertaken in different social, economic and physical environments. My treatment is not to illustrate agricultural regions or types of farming as such, but simply to exemplify the contrasts in the organising and practising of agriculture in these different areas. The first of these chapters 'Mixed Agriculture', deals with farming as carried on through a system of private enterprise and medium-sized or even small farms operated mainly by family labour. Many of the farms of the U.S.S.R., described in the next chapter, are also 'mixed', with both crops and livestock being produced on them, but the system of organisation and scale of operations is totally different, illustrating an entirely different approach to the production task. Hence, these chapters should not be seen as designed to fit exactly into the historic classification of world types of agriculture preceding them — they are not illustrative of 'types of agriculture' in this sense. They do, however, illustrate the framework within which types of agriculture and regions may be discerned. Similarly with the following chapters on plantation and subsistence agriculture.

Part III is concerned with the concepts and methodology that have been developed by geographers and other scientists for the greater understanding of any area or type of farming. This treatment is placed last because it is not essential to the understanding of the previous chapters but must precede more advanced study. Regionalisation and classification are considered, and again there is some detail to enable the study to be pursued without the benefit of the original works, to which, however, the advanced student or research worker must turn for further details. Classification of land use and land potential studies are dealt with in Chapter 10. These are subjects of particular importance for the improvement of farming and increasing of production, because those who create and develop policy in economic, social and political terms need the best information that can be provided as a result of academic sifting of material. Volumes could now be filled with the many statistical and other procedures developed by research workers, but only a few can be dealt with here, especially as the length of the book is being kept very close to that of the first edition in order to keep the price as low as possible.

The approach has, therefore, continued to be concentrated on basic and well-tried techniques which are described in reasonable detail with briefer references to other devices which may be more exciting methodologically but of less proven value. Some emphasis has been placed on techniques suitable for use in field classes, especially in land use and land potential evaluation.

Since the publication of the first edition there has been much more widespread attention than formerly to the vital question of feeding the world's population and to conservation of the environment. It is dismaying that, in spite of intense scientific work and economic support leading to substantially increased yields of crops, any long-term balance between world food supplies and demands seems to be becoming less and less likely. The rapid growth of world population and the effects of monetary inflation have, in fact, worsened the position of many of the less wealthy countries. There has been little liberalisation of world trade and, indeed, many see a distinctly counter-productive element in the enlargement of world trading blocs such as the European Economic Community, which restricts the import of food from more efficient agricultural systems, while increasing its own problems caused by the accumulation of surpluses at prices which place them beyond the reach of the mass of the consumers. If this led to diversion of food to more needy peoples this would be beneficial, but unhappily there remains no indication of a solution to the dietary and financial problems that impede such a realignment of trade in agricultural products.

Environmental deterioration is closely linked with agricultural problems. Not only in soil erosion but in pollution of rivers, lakes, seas and the atmosphere there are severe and increasing threats to agricultural production. The sterilisation of ever-increasing areas of land by new urban and industrial buildings and the failure to clear blighted urban areas — the legacy of past industrial ages — adds to the problems. Society, it seems, can find endless resources for advertising, commercial exploitation and non-productive activities, but cannot organise either production and distribution of essential commodities without gross ineffi-ciency and waste, or the protection of irreplaceable natural assets for the future.

Thus, the concern with man's treatment of his planet and his fellow men (which we are tempted to call 'inhuman' but is, regrettably, essentially human) that guided, in some part, this particular approach to agricultural geography, has in recent years become more acute, sharpening impatience with politi-cians, merchants and at least some scientists and academics for their lack of serious attention to these fundamental problems. Perhaps by the time the next revision of this book is needed, people will have adopted new attitudes and attempted seriously to reduce waste and to adopt practices in day-to-day living which will give more hope for the future, but it is not easy to be optimistic.

Acknowledgements

My first acknowledgement must be to Professor R. O. Buchanan for encouragement to write this book and for reading and commenting on the whole of the original script. Moreover, to him and to the late Sir Dudley Stamp I owe my own introduction to the systematic study of agricultural geography and land utilisation at the London School of Economics. I am also indebted to Associate Professor Peter Irwin, University of Newcastle, N.S.W., for reading the text of the revised edition and making many helpful comments. Many others deserve mention but it would be impracticable to acknowledge all the help and ideas that have contributed to this book. To all I am deeply grateful, though responsibility for any errors and all omissions is entirely my own.

I am pleased to acknowledge the co-operation of the following authorities for permitting me to use their material: the Soil Survey of England and Wales, and U.S. Department of Agriculture (Figure 1); Messrs George Allen and Unwin Ltd. (Figure 2), Messrs Hodder and Stoughton Ltd. (Figure 14), The Association of American Geographers (Figure 5), The Editor, *Journal of Tropical Geography* (Figure 11), The Editor, *Economic Geography* (Figures 12, 13, 15 and 16) and Messrs Routledge and Kegan Paul Ltd. (Figure 10 and associated material). Extracts from maps of the Land Utilisation Survey of Great Britain and the second Land Utilisation Survey of England are reproduced in Figure 17 by permission of the late Sir Dudley Stamp and Miss Alice Coleman respectively. The sources of other maps indicate the authorities who have kindly given permission for their use in this volume.

I wish to thank the several cartographers and photographers who provided technical services. For this edition Mrs G. Bridges drew the revised Figure 1 and Mr G. E. Lewis the new Figures 4, 8 and 9. My wife devoted many hours to preparing the index, proof-reading and other tasks.

Introduction

Agricultural geography is the study of agricultural activities in a spatial context, that is, the location of agriculture as a whole and of the constituent activities of cropping and livestock rearing, and the distribution of the outputs (crops and livestock) and of the farms, fields, labour, machinery and all other inputs required for production. It is one aspect of the study of geography which is concerned with the spatial patterns of human and physical phenomena and their inter-relationships.

Agricultural geographers do not look upon their study as constituting a major divison of geography but as a part of economic geography, which is regarded by some authorities as a discipline in its own right, or at least a primary division of geography.[1] More widely, however, it is considered as a part of human geography, which alone is placed with physical geography to constitute two primary divisions. Many definitions of economic geography have been attempted, but none has the merit of combining adequacy and brevity better than 'the geography of man's work'.[2] Agricultural geography fits into this definition as the geography of man's husbandry of the land, exercised through the tending of crops and livestock for subsistence or for economic gain.

Hunting and collecting are not here regarded as agriculture unless there is conscious effort to improve, or at least modify, the hunting or collecting grounds beyond their natural state. Modification of the natural environment is the essence of agriculture. It may be argued that every hunter or collector modifies the environment by limiting the spread of his prey, be it animal or fruit, but a reasonable interpretation of agriculture

[1] See for example the Russian view expressed by Konstantinov (1962).
[2] Buchanan (1951).

1

would seem to be deliberate effort to modify. Thus, when the hunter deliberately burns the range or selectively culls animals, he is on the verge of becoming a pastoralist, and for practical purposes we should include pastoralism in agriculture. When man follows fire by scattering seeds of desired species in the ashes, he is an agriculturalist, even if he does not so much as turn the soil with a digging stick.

Whether or not to include forestry in agriculture is a question not susceptible of easy solution. In parts of Scandinavia and central Europe, for example, forestry is an important part of the agricultural economy. Forest lots form an integral part of the overall and long-term management of the individual farm. Even so, the day-to-day integration of use of the forest and farm is usually limited to pasturing stock in the woodland, and this only where it can be done without serious harm to the trees. More generally, farmers see forests as competitive rather than complementary with their activities, though scientists and foresters may interpret the relationships otherwise.[1] On balance, and having regard to the moderate length imposed on this book, I have excluded forestry from direct consideration.

Modern economic geography is not merely the geography of production or even of production, exchange and consumption. Geographers have become increasingly concerned with social problems, and economic geography has evolved from being mainly descriptive to a more mature concern with the location and social significance of the conditions of human societies and their behavioural patterns, employing fairly sophisticated quantitative techniques.

There has been an immense change in the way geographers, along with other scientists, have viewed the impact of the physical environment on economic and social activities. Some decades ago 'determinism' expressed the belief that the natural environment closely controlled man's activities, to the extent that the organisation of his life was a response to the restrictions imposed on him by relief, soils, climate and other manifestations of the physical surroundings. Revolt against the acceptance of such limitations produced the 'possibilist' school of thought, which holds that the environment offers certain possibilities which man may or may

---

[1] See for example 'Agriculture and forestry; competition or coexistence', *International journal of agrarian affairs*, II, 1955.

not exploit. Determinism is generally discredited but 'probabil-
ism' offers a compromise position.[1]

In much of man's life today the physical environment does not
greatly constrain activity. Except at the absolute extremes people
pursue very similar occupations and dwell in remarkably similar
buildings in all parts of the developed world, and other regions
are fast developing similar life styles. A man may proceed from a
very similar house by an almost identical form of transport to an
exactly similar job in cold Scandinavia, cool and humid British
Columbia or warm and dry Australia. Even such distinctive
cultures as those of Japan and China have not prevented the
adoption of similar artifacts, even to personal clothing and
high-rise buildings, to those which have become commonplace in
the western world. Differences of detail remain, but in most fields
of activity they are variations on the theme rather than the theme
itself that the physical environment encourages.

In agriculture, however, the environment still has a greater and
more direct effect. It is true that many aspects of farming are very
similar on greatly contrasting soils in very different climates.
Dairy farms in New Zealand and Denmark may keep the same
breeds but the organisation of work required to produce the same
product will be very different. An effective illustration of the
contrasting extent to which man may conquer his physical
environment in agriculture as compared with other activities is to
be obtained by reference to the U.S.S.R. With one-sixth of the
earth's surface under unified political and economic control, and
with a philosophy which has stressed man's ability to dominate his
environment pursued by the Soviet rulers for over half a century,
an immense contrast is seen in their successes in this direction in
agricultural and in other fields.

In factory employment, as in the standardised forms of
housing and retail establishments, transport media, educational
systems and health and welfare services, there is little sharply
different between north-west Russia and the Ukraine. But the
pattern of agriculture still shows a healthy respect for variations
in the physical environment, and this is not for want of effort to
overcome this dominance and reduce specialisation. Thus, N. S.
Khrushchev, when directing the country's affairs, decreed that

[1] Space precludes more than brief reference to this philosophical argument.
Lewthwaite (1966) provides a readable guide to the debate.

maize should be grown much more widely throughout the land and the area sown to maize increased from 6 million to about 37 million hectares between 1953 and 1962, but the failures of the crop in cool and moist regions led to retrenchment.[1] Reduction of the maize area was not to the old figure but to an intermediate one, averaging 23 million hectares in 1965–69 and 21 million hectares in 1973, representing the kind of compromise that has to be reached in practical farming, in which yields, costs and prices cannot be ignored even in a state-directed socialist agriculture. As between any European part of the U.S.S.R. and Soviet Central Asia, the agricultural contrasts are even more sharply preserved.

Within the smaller areas, such as in differentiation of farming in England, or still more, within a region such as East Anglia, the investigator may well emphasise the non-physical parameters, but on the larger canvas the major variations still reflect physical conditions. Maize or any other crop still grows better in an appropriate soil with adequate sun and moisture than in less suitable land and, other things being equal, the product will be cheaper. Not only climate but the daily variations of weather constantly affect farm planning and operations and bad weather frequently brings serious losses for farmers[2] and, ultimately, higher prices for consumers.

Hence, it is apparent that climate and weather set broad limits to the types of agriculture that man may pursue in any region. Within these limits, relief factors impose further limitations or, to express the same idea in a positive instead of negative way, offer certain possibilities. The relief controls operate partly through the climate they create, high altitudes being in some respects similar to high latitudes, and partly through angle of slope and irregularity of surface. The soil reflects both these major factors (climate and relief) as well as the vegetation, which is itself an expression of the interaction of climate, relief and soil. Also in the character of soil, partly through his effect on the vegetation, but also directly, is found the impact of man.

Man modifies his environment continuously, both consciously and unconsciously. The way in which he makes these modifications and adaptations depends in large part on the economic results which he perceives as likely to result from his efforts — the

[1] Anderson (1967).
[2] Taylor (ed.) (1970) provides numerous examples.

revenue he will get for his products, the prices he must pay for the goods and services he wishes to purchase, and the cost he incurs in production. Cost is not to be evaluated solely in terms of monetary outlay, for, as the economist teaches, the cost of something is what is gone without to achieve that thing. The cost to a farmer of spraying his wheat to kill a particular insect may be the improvement of housing to protect valuable livestock in winter — the cost of not spraying may be the loss of the wheat. Unfortunately, true costs of this kind are difficult to evaluate, but they must influence the policy of the individual and of governments.

Government policy will itself be a major factor in fashioning the economic environment for the farmer. Also conditioning his response, and, indeed, the range of choices open to him, will be the services at his command — scientific, technological, financial and others. These, and the individual's initiative in making use of them, will be affected by the historical and cultural background of his nation and local community and by his religious and other personal commitments.

The accumulated effects of the physical, social and economic factors from the past have become merged in what is now commonly called the historical factor. By this is meant the sum total of the influencing of present patterns of behaviour by what has gone before. Owing to the limitations of length in this book, together with the fact that the study of past distributions can be more effectively treated in works of historical geography, the book will not pursue any detailed study of historical influences, although some effects will be apparent in several chapters. However, the diffusion of ideas, implements and methods of agriculture must be regarded as of fundamental importance because knowledge must precede adoption. The understanding of the antecedents of present-day agriculture is, like all knowledge, derived from the workers of many disciplines, with geographers particularly contributing to the research into paths and means of diffusion. This is the major theme in the brief historical resume which follows immediately, before the chapters on physical and economic factors.

PART I

THE PHYSICAL AND SOCIAL ENVIRONMENT

CHAPTER 1

The Origins and Development of Agriculture

Every agricultural landscape is made up of individual farms, and it is the decisions of the owners, managers and workers of all those farms which result in the productive pattern that produces the cultivated landscape. The total farming landscape depends also for its appearance and functioning on the work of non-farming sections of the community in the transport network, power, drainage, woodlands, factories, shops and houses. All these elements have been constructed and modified by individuals and communities in response to the needs and pressures of their times, making use of the knowledge and artifacts that they have inherited from the past and those that they, their neighbours and others produce in their own lifetimes. The process of diffusion of knowledge and techniques is essential for progress or even to maintain stability in the face of invasions of pests and diseases and other threats to the existing way of life.

Any planner, administrator, or other reformer who enters the landscape with the intention or wish to influence its development, or merely to understand its functioning or interpret it for others, ignores the past at the risk of failing in his task and, if it be a task with a practical result, of disrupting and disorganising the community in the area and leaving it in a worse state than before.

Just as the individual farm is the basic element of the agricultural landscape or region that should desirably be studied to understand the present economic functioning of the area, so it is the individual farm's history that should be analysed to arrive at the background of an existing pattern. In most cases, however, this is impracticable. Indeed, every farm is the successor of innumerable past farms, which have either disappeared completely or have left only slight traces of their former existence in field patterns, trees and hedgerows, ditches or buildings. Furthermore, even if numbers did not make it impossible to analyse every farm individually, concern for the privacy of the individual largely rules out such detailed investigation. So, with such sampling of individuals as may appear appropriate, the researcher turns to the more general background of crops, livestock, artifacts, labour, management and capital to explain the regional pattern and understand its needs.

How far back one should go into the past to explain the present is not capable of generalisation. For practical purposes today it is not so necessary to know that specialists trace the origins of wheat and barley to the Near East some 12,000 years ago, as to know the range of varieties now available from crop breeders and seed distributors. But for the intellectual satisfaction of historians, historical geographers and botanists, archaeologists and anyone involved in the history of agriculture there can be no limits to research. Somewhere along the line the purely academic merges with the practical and becomes relevant to present needs and the future. As to where this happens, again no generalisation can be offered, and every case must be judged on its merits. It can only be reiterated that it is folly to try to understand any region without considering its past, as may be seen on reflecting on many of the cases that are mentioned later in this book, such as the problems of productivity in Russia, of farm size in Ireland, of specialisation in New Zealand for a distant market, or of product selection in Malaysia. Every researcher should, therefore, have a broad knowledge of the history of agriculture and the diffusion and development of its constituent parts and know where to pursue greater detail when and where needed. Here it is possible only to provide introductory notes on agricultural history as some background to the later chapters. More detailed surveys are readily available.[1] Before man's earliest attempts to control or influence vegetation, plants provided him with some food directly and some through the insects, animals and birds they supported. Use of fire-culture in some deliberate form, such as clearing forests and driving game, may take us back hundreds of thousands of years, but almost certainly tens of thousands.[2] The

[1] These include Ucko and Dimbleby (eds) (1969), Thomas (ed.) (1956), Clarke (1952).
[2] Stewart (1956).

setting of fire to improve pasture is both old and widespread, and selection of plants, some to be encouraged, some to be destroyed by fire and other means, undoubtedly preceded the beginnings of agriculture proper.

The domestication of plants and animals probably began rather later than 10,000 B.C.[1] Early domestications included wheat and barley, sheep, goats and, later, cattle and pigs. Vegetative reproduction was probably understood earlier and utilised for food production in south-east Asia but archaeological evidence is much less readily available than for seed cultivation. Animal domestication may have originated with either sedentary cultivators or nomadic herders. Pastoralism may have developed from the modification of mixed farming as this was taken into the arid areas of central Asia, or mixed farming may have originated from the merging of seed cultivation and stock herding, as differing economies merged in the region of the Fertile Crescent.[2]

Agriculture is more likely to have developed with the economically more generalised gatherer-hunter-fisher populations occupying the more varied natural ecosystems, such as the forest and woodland margins, where a great variety of plants and animals were available, than with hunters occupying specialised ecosystems with limited species such as the natural grasslands.[3]

To Sauer,[4] the most likely people to begin agricultural practices were 'some well-situated, progressive fishing folk living in a mild climate along fresh waters'. He proposed south-east Asia as the cradle of earliest agriculture. 'No other area is equally well situated or equally well furnished for the rise of a fishing-farming culture.' He argued that the earliest domesticated animals — dog, pig, fowl, duck and goose — originated there as animals of the household, in contrast to the herd animals of south-west Asia. Here also is the major centre of planting techniques and vegetative reproduction which, he accepted, man learnt before he learnt the growing of crops from seeds.

Diffusion of these cultures, in Sauer's view, occurred in the Pacific region, northwards to China, round the Indian Ocean to Africa and through the Mediterranean lands to Europe. Similar

[1] Ucko and Dimbleby (1969).   [3] Harris (1969).
[2] Narr (1956).   [4] Sauer (1952), 23–29.

reliance on vegetative reproduction is found in tropical America. He favoured the north-western extremity of vegetative planting, that is, the Mexican-Central American border, as the hearth of the principal seed plants. In the Old World, Sauer recognised three centres of seed domestication — one in north China, a second in western India, extending to the eastern Mediterranean and a third in Ethiopa. 'In all three, vegetative reproduction was made difficult, annual seed growing facilitated by climate.'[1]

Earlier, Vavilov[2] had listed eight independent centres of origin of the world's most important cultivated plants, based on expeditions he and other Russian scientists made throughout the world between 1916 and 1934. These included China, which he regarded as the earliest and largest centre, India with Burma and Assam, Malaysia, Indonesia and the Philippines, which total area contributed some 300 species including rice and millet, soya beans, sugar cane, bananas, coconuts and other fruits and vegetables. Vavilov thought Central Asia, including north-west India, the home of some grains, legumes and cotton but rightly found south-west Asia more important for wheat and rye, alfalfa, temperate fruit and nuts. He also believed that some grains originated in Ethiopia. South Mexico and Central America were credited with the domestication of maize and the sweet potato and the Andean region with the potato. He noted the Mediterranean area as particularly important as a secondary source in which much selection took place as well as being the area in which the olive, vine and fig were domesticated.

More recent scientific work has modified these pioneering views. The general picture as summarised here is substantially unaltered but more importance is assigned to south-east Asia and Africa south of the Sahara and less to China, India and Central Asia. In each of the old world hearths of seed agriculture it would seem that cultivation included grasses for grain, legumes for protein and fat, and usually some additional oil and perhaps fibre plants. The distinctiveness of the agricultural complex that developed between the eastern Mediterranean and south-central Persia is the combination of seed growing with herding of animals — sheep, goats and cattle — in which we find the origins of modern mixed farming.

This early mixed farming of the Near East was evolving

[1] Sauer (1952), 72–73.   [2] Vavilov (1935, 1949–50).

between 9000 and 5000 B.C. The first agriculture in Europe has been identified as existing before 5000 B.C. in Greece with widespread diffusion in the following millennium, to reach Holland before 4000 B.C. The sites of southern and central Europe reveal economies based on the cultivation on easily-worked loess soils, of wheats, barley, beans, peas, lentils and flax, and in some areas stock raising included oxen, sheep and pigs.[1] Cultivation was by primitive hoes, and was probably of a shifting kind, if only because of the lack of manure. In the later Neolithic economy of northern Germany and southern Scandinavia, however, cattle played an important part. Settlements became more permanent.[2] Clearance of the forests was beginning to outstrip their regenerative capacity, a condition that extended to the Rhineland and the Netherlands by the beginning of the Bronze Age. Throughout the Neolithic Age the diffusion of agriculture continued and settlements dated to as early as 2500 B.C. show that the grain and animal economy had extended to Ireland.

The Bronze Age, approximately 1500 to 500 B.C., saw the spread thoughout northern Europe of a light plough, the ard, which would not have been capable of breaking in new ground but would have permitted more efficient tillage of land already in cultivation. The succeeding Iron Age was the period in which the heavier soils came under cultivation with the development of the iron ploughshare.[3] Iron tools did not begin to affect cultivation in Britain until about 300 B.C., but were part of the civilisation of southern Europe by 1000 B.C.

About 800 B.C. the Greek *polis* or city settlement was developing, based on the improved agriculture possible with the use of iron. Hills were terraced for the vine and olive, and these fruits together with grains and Greek breeds of the principal domestic animals were sent to the colonies, which extended from Egypt and the Crimea to the Iberian peninsula. In return Greece acquired other plants and animals, including the domestic fowl from the East. By the fourth century B.C. Greek agriculture and its administration were treated as a fine art. During the following centuries many agricultural treatises appeared and the science acquired its Roman name.

[1] Murray (1970) summarises the evidence.   [3] Evans (1956).
[2] Clarke (1952), 97.

During the period of Greek and Roman colonisation and supremacy, agricultural products and ideas were exchanged throughout the great region of their influence, and beyond it through intermediate traders. New scientific devices, such as the Archimedes screw, made possible new approaches to irrigation and other aspects of land improvement and cultivation. All this development was stimulated by the growth of a monetary economy and the abandonment of subsistence agriculture in favour of production at least partly for a market. Estates organised by government agencies became dominant in large areas of the classical world. The *latifundia*, great estates based on slave labour, took the place of peasant settlements in much of the Roman world. As the supply of slaves diminished in the changed conditions of the *pax Romana* of later centuries, there was some reversion of organisation of these estates to a type of peasant tenantry, and subsistence farming became again more common.[1]

The second century A.D., however, in spite of a decline in farming for the market, saw the spread of techniques, tools, plants and domestic animals beyond the Roman world as far as Ireland, Scandinavia and western Siberia. As in Rome in the fourth and fifth centuries, the feudal system bound the peasant or serf to the estate in much of Europe. In the Byzantine east, freeing of the peasantry began in the sixth century, but this movement was not paralleled in western Europe. The monasteries which developed in Egypt in the fourth century were, however, followed in the west during the succeeding centuries by the growth of the strong religious communities which kept alive scientific agriculture and land development.

THE MIDDLE AGES

In the millennium that preceded the agrarian revolution of the eighteenth century and the development of commercial farming as we know it today, developments in the techniques and administration of agriculture came at long intervals. The feudal system, in which the vassal owed allegiance to his lord in return for a measure of security, was expressed on the land by manorial organisation. The vassals supplied the lord with a portion of their produce and also had to work on the desmesne lands farmed for the lord's benefit. The system was partly a response to an

[1] Heichelheim (1956).

economy in which money was scarce, as in western Europe, which was drained of gold and silver by the shift eastward of economic and military power, and largely cut off from the Mediterranean, Africa and the east by Arab conquests. The feudal and manorial systems spread throughout the Carolingian Empire and after the Norman conquest of England the military obligations of the feudal system were grafted on to the manorial system already developed by the Saxons. Even in the Carolingian Empire the manorial system varied regionally[1] while Scandinavia and the North Sea coast remained little affected by it, having a trading economy based on stock rearing.

Field shape and utilisation also varied more than was formerly supposed. The open fields, which were not fenced individually and were cultivated in strips,[2] have become identified particularly with the Middle Ages, but in some areas individually-walled fields survived, and in regions of colonisation such as the fens of Holland farmhouses were sited on the individual strips.

Five methods of cultivation have been distinguished[3]:

(1) Temporary cultivation, after which the land was allowed to revert to waste for an indefinite period.

(2) The infield-outfield system of Scotland and Ireland and similar systems in which part of the land was cultivated continuously with heavy manuring, and other parts occasionally.

(3) The two-course rotation, in which the land was tilled and left fallow in alternate years.

(4) A three-year system in which the land was tilled for one year and left fallow for two.

(5) The three-course rotation, in which winter corn (wheat or rye) was followed by spring corn (barley or oats) and then by fallow.

Two-course rotations were practised mainly in the Mediterranean area where winter corn utilised the seasonal rainfall, and in northern Europe where the seed bed could not be prepared until spring. In the intermediate zone both winter and spring grains could be grown. The three-course system offered increased

---

[1] Slicher van Bath (1963).
[2] Beresford and St. Joseph (1958) provide an attractive survey.
[3] Slicher van Bath (1963), 58–59.

production but not sufficient to meet the needs of a rapidly expanding population.

In the late Middle Ages (the second half of the twelfth century and the thirteenth century) rising population and growing circulation of money led to a great increase in cereal prices which stimulated reclamation of marshes and forests and ploughing-up of pastures. As in later periods of population pressure, colonisation extended into marginal lands which, in the absence of adequate manure, soon lost their accrued fertility. Farms on the more fertile lands were subdivided until in many areas they became too small to support the people dependent on them. The economic change of this period and the depression that followed in the fourteenth century led to the gradual abandonment of the manorial system. Villeins were able to buy their freedom and in some cases were compelled to do so. A new class of small tenant farmers and cottars became widespread.

From 1300, wheat prices fell steadily. Plagues, including the Black Death of 1347–57, swept Europe, and the depression deepened in the fifteenth century. Cropping was reduced and livestock increased, sheep being especially favoured to meet the growing demand for wool. Even before the end of the twelfth century many of the Cistercian farms, their superior organisation facilitating improvements,[1] were specialising in sheep rearing, but it was in the fifteenth century that the widespread adaptation of agriculture to commercial farming occurred. Industrial crops, including hops, flax, hemp, dye-plants and oil-seeds, became more important and more vineyards were planted.

In the sixteenth century cereal prices rose again, reflecting new expansion of the population and increased use of horses. A new wave of reclamation took place and the search for higher yields, with intensified manuring, extended to both food and industrial crops. The colonisation of tropical and sub-tropical areas led to the invention of the plantation system for commercial crop growing.

There were periods of depression in the seventeenth and eighteenth centuries and fluctuations in emphasis on arable and pastoral husbandry according to their relative prosperity, but the general growth of population and the improvement of communications were setting the stage for the new husbandry which began

[1] Donkin (1963).

to spread across Europe in the late eighteenth century. The transition to the new methods was slow and often indirect. The replacement of methods involving fallow by continuous rotations, including root crops, took place in stages which varied from place to place and was spread out over several centuries. Slicher van Bath lists eleven distinct tillage systems involving varying periods of fallow or fodder crops found in the seventeenth and eighteenth centuries.[1] Turnips, rapes and clovers were already an important part of the farming systems of the Low Countries.

THE AGRARIAN REVOLUTION

The agrarian revolution was intimately interwoven with and dependent on the industrial revolution. As has often been pointed out, these were no sudden revolutions, but complexes of slowly evolving technological improvements which became gradually more widely disseminated. Thus, although Abraham Darby was smelting iron with coke at Coalbrookdale by 1709, similar methods were not adopted outside the Coalbrookdale and Wrexham districts until after 1750. Many improvements had to be incorporated before coke smelting could be used for all types of iron but as the new methods in iron-working became more widely adopted iron became cheaper and more readily available to other industries, including agriculture. Steam power became generally available, and was put to work in farmyard as well as in factory. The construction of canals, and later railways, linked the factories to their raw materials and provided the transport for agricultural produce to the growing urban markets. The factories and ancillary services grew apace and their demand for labour was insatiable. Labour was drawn away from the land, and although this resulted in some labour shortage, it stimulated attention to more efficient agricultural methods. Above all, it facilitated the re-arrangement and enclosure of fields that were a vital prerequisite of the agrarian improvements.

Specialisation was an important feature of the new agricultural methods. Limited by primitive transport and commercial services as well as by traditional techniques, most farmers had little notion of producing a surplus for sale before factories, mines and transport networks had created a landless working force which had to be fed by a surplus from the land. Subsistence agriculture

[1] Slicher van Bath (1963), 244.

satisfied the requirements of the cultivators, more or less, for food, drink and clothing. A particular crop, such as wheat, in regions where it was favoured, was produced on land little suited to it, as well as on suitable land. Yields were low, and recuperation of the soils under fallow was inadequate. The introduction of roots not only enabled fertility to be better maintained, while yielding extra crops, but facilitated a degree of specialisation, as between turnips for stock and potatoes for human consumption.

In the arable fields continuous cultivation to keep down weed-growth was made possible by Jethro Tull's drilling and horse-hoeing husbandry, whereby seed was sown in straight lines by drills so that inter-row cultivation was possible. This was incorporated with the principle of alternating exhaustive and recuperative land uses in the Norfolk rotation. Temporary grass followed spring corn (barley or oats), and was ploughed-in, to be followed by winter corn (wheat), and finally roots, to complete the rotation. It was later found necessary to winter large numbers of stock in order to manure the land. This became the standard system of the light soils in the eastern parts of England.[1]

More important for the wetter districts was the harnessing of the clovers and related plants to improve the nitrogen cycle in the grasslands. The ryegrass-clover sward and its variations provided the basis of improved pasture and also meadows which yielded hay sufficient to keep livestock through the winter in increased numbers and improved health, so permitting an increasing supply of meat and dairy produce from the regions not suited to arable farming.

The regional variations in land use that developed in the eighteenth and nineteenth centuries could exploit more fully the potential of the European climates because of the improved breeding of livestock. In cattle breeding, as horses became more widely preferred to oxen for work in the fields, emphasis was placed increasingly on selection for meat and milking qualities. Further adaptation to local climatic, soil and transport conditions led to refinement of breeding to improve either the milking or the fattening qualities, so producing the specialised dairy breeds such as the Jersey, Friesian, Ayrshire and Dairy Shorthorn and the beef breeds, Hereford, Aberdeen-Angus and Beef or Scotch Shorthorn. Sheep breeds, similarly, were improved by selective

[1] Smith (1949), 23.

breeding, to serve the needs of the manufacturers for different grades of wool, and of butchers for different classes of meat. At the same time, the more specialised breeds were able to exploit more efficiently the varying regional environmental conditions, so encouraging further specialisation on the individual farm.

The improved agricultural systems did not become generally adopted in all districts and by all farmers, even in the British Isles, until not merely decades but a century or more had elapsed but by the beginning of the nineteenth century 'the new order was dominant and the old recessive'.[1] The farming of the lowlands had then been largely brought into line with the new techniques, which were being carried rapidly into the more remote and less productive hill regions.

Elsewhere, diffusion carried other systems of exploitation into hitherto isolated areas. In North America, for example, cattle ranching, developed in south-east Texas by southern Anglo-Americans between 1820 and 1840,[2] was thereafter diffused throughout Texas and ultimately throughout western America. Sheep rearing, found to be an effective way of utilising the semi-arid lands of Australia, was carried on across the Tasman Sea in the 1840s and was diffused throughout New Zealand in the following decades. Innovations in grain farming developed in America were adopted also in Australia, and specialisation emerged in all the 'new' areas to meet the challenge of growing markets as the improving transport of the age of steam made available in Europe the produce of the new lands, particularly the Americas. By adaptation of European methods to exploit the large areas of land available, utilising less labour and more machinery, costs of production were cut to the point where grain could be shipped and marketed in Britain more cheaply than the local produce. The repeal of the Corn Laws in 1846 in response to manufacturers' desires to keep down costs and, therefore, wages, had opened the way for the import of greatly increased supplies of food as these became available. By the 1870s the years of general agricultural prosperity in Britain were giving way to changing conditions and low prices for grain spread to livestock products as, from 1882, refrigeration made it possible for meat and dairy produce from New Zealand, Australia and South

[1] Smith (1949), 44.
[2] Jordan (1969).

America to be marketed in Britain. During the 1870s and 1880s poor seasons and livestock diseases contributed to the problems of farming in Britain but the depression speeded up the adoption of innovations as the more enlightened farmers sought higher productivity.[1]

One of the most important developments of the second half of the nineteenth century was the increase in use of fertilisers and the proliferation of factories to manufacture them. Increased manuring and marling of land had, of course, been part of the earlier agricultural revolution, and a major force behind the enclosure movement had been the desire of the improver to benefit from increased fertility, but manufactured fertilisers, especially nitrogenous compounds and superphosphate, reduced the dependence of the arable farmer on livestock and enabled the livestock specialist to keep many more productive animals per hectare. The breeding of new varieties of crops and livestock contributed to increased productivity and fuller benefits from all developments came with the refinement of chemicals for pest and weed control and reduction of diseases affecting plants and animals. Technical progress was accompanied by increased attention to farm economics and the appreciation of the importance of the balance sheet was a feature of the new generation of farm managers, leading, incidentally, to the foundations of the systematic study of the geographical location of agricultural activities.[2]

Farming in Britain, adjusting to free trade, became more efficient with more economical use of labour, machinery and fertilisers and amalgamation of farms to form more economic units, but in most parts of Europe the revival of protectionism to preserve high prices for the farmer — and for the consumer — shielded the agricultural industry and delayed modernisation.

As improvements in transport spread, however, backward agricultural regions experienced some of the changes initiated elsewhere. Exports of grain from Russia rose from 5 per cent to over 14 per cent of the gross grain harvest between 1860 and the late 1870s.[3] Following their emancipation from serfdom the peasants were forced to sell more on the market to meet their

[1] Perry (ed.) (1973).
[2] The contribution of Von Thünen and later workers is reviewed in Chapter 9.
[3] White (1975), 12.

monetary obligations and the railways provided the means. A marked increase in specialisation of area in grain production followed the transport improvements as production was concentrated in the higher yielding areas.

For the first quarter of the twentieth century farming continued to evolve on what might be called the patterns of the nineteenth century, both in terms of land use and trade. The First World War set the scene for much of the change that was to follow in technical, commercial and political developments.

A decade after the revolution that gave the Bolshevik Party power in Russia, the U.S.S.R. embarked on agricultural collectivisation. Private ownership of land having already been abolished, the peasants were forced into collective or state farms, an experiment followed in varying degree by other countries in which communists have gained control, so bringing ultimate management by the state to much of the world's land and agricultural labour force. In other countries, government has increasingly exerted influence over farmers through price controls, market organisation, production subsidies, purchase guarantees, health regulations and innumerable other ways. All of these have grown rapidly since the Second World War, as have international agreements intended to limit competition and provide security for producers, though this has meant shielding the less efficient as well as the more efficient among those protected.

Agricultural productivity has been enhanced by the many technological innovations that have come from mechanical and civil engineering, botany, biology, chemistry and other sciences. Perhaps the greatest single agent of change has been the internal combustion engine, which revolutionised the application of energy to the land. Between 1925 and 1950 the tractor replaced animal haulage on the more advanced farms of the world and during the next quarter-century it completed its dominance of the cultivation and farm transport in the western world and the U.S.S.R. and began significantly to contribute to production in the Third World. In some aspects of production, electricity has been no less an agent for change, particularly in production per man in milking, shearing and other livestock operations. Between them these energy forms have facilitated the spread of irrigation and drainage networks, the application of fertilisers, the control of pests and diseases, the processing of produce and, indeed, almost every part of farming. Even the division of the land into fields has been changed as the nineteenth-century invention of barbed wire has been supplemented and, in some cases, replaced by the movable electric fence.

During the same half century, from tentative beginnings in 1922 in the U.S.A. and the U.S.S.R., the application of aviation to agriculture has become a significant aid to farming in almost every country of the world. Hitherto used mainly to deal with invasions of locusts and other special situations, the aeroplane became in the 1950s in New Zealand a routine tool for applying fertiliser and a whole new contracting industry developed there, using firstly British war-surplus, then American specialised aircraft and ultimately home-produced designs. In the U.S.A. and U.S.S.R. the spraying and dusting of millions of hectares became an annual programme in the 1960s while the 'seventies have seen similar operations become routine in the Third World, with contracting aircraft moving seasonally from Europe and the U.S.S.R. to Africa and Asia and back.

Technological advance has not only brought gains; the negative aspects have also multiplied. In particular, the potential for serious damage to the environment has increased, and agriculture cannot be exonerated from blame for some of mankind's foolish or misguided acts. The lessons from creation of the 'dustbowl' in the midwest of the U.S.A. in the 1930s, through extending cultivation into too-arid areas, had to be relearnt in the Soviet virgin lands in the late 1950s; chemicalisation of the land, though necessary for optimum production, gives rise to many misgivings, which are increased by the speed of application from the air of pesticides, fungicides, herbicides and fertilisers; irrigation in arid areas poses the problem of secondary salinification. Many other examples of undesired effects from development could be cited.

Furthermore, the spectacular technological advances and the applications of capital investment, especially in high-energy applications, increase, rather than diminish, human responsibility on the farm. Though the demand for direct labour has been reduced, day-to-day operations in cultivation and livestock husbandry, as well as management, require more training in a wider range of skills than ever before.

Neither has the availability of increased knowledge and the refinements of seeds, fertilisers, machinery and other inputs and the widening of control and investment by governments diminished the significance of physical factors. More is known about the overcoming of limitations of soil and climate, but on better land costs are still lower per unit produced and rents can still reflect this difference. The occurrence of droughts still results in severely lowered yields, not only in areas where agriculture is still primitive and remedial measures of limited availability, but also in areas such as the U.S.S.R., where vast state resources can be mustered, and in highly developed capitalist societies where personal involvement and managerial skills have been able only to mitigate, not eliminate, the consequences of climatic fluctuation. This is well illustrated in the droughts of the U.S.S.R. in 1975 and in western Europe in 1976. It is thus appropriate at this stage to examine in greater detail the individual factors and forces that make up the physical environment which agriculture exploits and with which it has to contend.

CHAPTER 2

Physical Factors influencing Agriculture

The physical factors will be considered under three main headings; climate, soil and relief.

CLIMATE

Climate is the principal aspect of the physical environment affecting agriculture. The characteristics of the soil, the essential medium for plant growth, are largely the product of present and past climates and the vegetation that has flourished in them, and the effects of relief are to no small degree expressed through resulting climatic variation.

Every form of plant or animal life requires certain conditions from its environment for it to be able to survive, and somewhat more stringent conditions if it is to reproduce naturally. Agricultural systems usually make use of only a small number of the economically valuable plants and animals that are suited to a given environment, the actual selection being in response to economic and social conditions, present or past.

Precipitation and Water Relationships

Agriculture makes use of water derived from the soil and underground water table for most plant cultivation and water collected from rainfall or drawn from rivers and streams for livestock. Wells also provide for livestock as well as for human consumption and any source of free-flowing fresh water may be used for irrigation.

Since plants must derive the bulk of their water requirements through their root systems in order to make use of it, water must be available in the soil in the quantities needed by the plant. Too little water in the soil will result in the withering of the plant, too much will cause waterlogging and disease or death of the plant. Consequently, the relationship between climate and the characteristics and condition of the soil are of maximum importance for

21

plant growth. The efficiency of precipitation is indicated for general purposes by the comparison of evaporation with precipitation, conveniently expressed as P/E. Given only precipitation records and measurements of evaporation this is a simple and convenient indication of the value of a climate for agriculture in terms of water availability. For the productivity of any particular soil, however, what really matters is how much water enters the soil, i.e. the level of infiltration and the extent to which it is retained in the soil. Infiltration varies according to the nature of a soil, especially its texture, its condition (whether compacted, cultivated, etc.), relief, and the amount and type of plant cover. It also varies according to intensity of rainfall and its duration, and temperature conditions. Infiltration levels below 50 per cent and above 90 per cent are encountered. Since infiltration approximates to rainfall less run-off, in order to measure infiltration it is necessary to record run-off as well as rainfall for each type of soil and vegetation cover. This presents considerable technical difficulties, which can be overcome for practical purposes by use of a variety of infiltrometers and other instruments.

Infiltration is usually much less beneath arable crops like corn, cotton and potatoes than beneath grass, trees or mulches, and this must influence cropping policy in areas where soil moisture is limited. The beneficial effects of protecting the soil by mulching when intensive tillage is undertaken in dry conditions is a lesson stressed by American soil conservation authorities.

While all plants need water in order to survive, the requirement varies. It depends on the extent to which structure protects the species from transpiration, since nearly all the additional water required by a growing plant is to replace its essential water content as this is lost by transpiration. Per tonne of dry matter produced, cereals transpire 400–500 tonnes of water, and grasses more than 800 tonnes, according to the evaporative power of the atmosphere. A field producing $7\frac{1}{2}$ tonnes of dry matter per hectare may transpire from 3000–6250 tonnes of water, equivalent to 1250–2500 mm of rain.[1]

Plants vary also in their ability to extract water from the soil. In favourable conditions the roots of lettuce and spinach penetrate only 30 to 40 cm, those of potatoes and peas about 60 cm, tomatoes and tobacco 90 cm, field corn and asparagus 120 cm,

[1] Watson and More (1962), 38.

and alfalfa (lucerne) and grapes down to $2\cdot5$ metres or more.[1] This indicates one of the reasons why in any given physical conditions the range of cultivable plants is limited. Within this range, the maximum utilisation of any given conditions can be achieved only with suitable crops. A classic example of this is the introduction of alfalfa to the Argentine pampas where the water table is $1\cdot5$ to 5 metres below the surface of the ground. This solved the problem of fodder production, which could not be met by unsuitable native grasses, and so made possible the development of the beef industry.

For animals, apart from the water needed for the growth of their food, direct water requirements are considerable. Dairy cows naturally have the greatest needs. A Friesian cow yielding 36 kg of milk per day may require about 90 kg (85 litres) of water for liquid consumption. A Jersey cow producing about 13 kg milk might consume about 45 kg of water. Beef cattle need about 30 kg of water when fattening, and about 15 kg for a maintenance ration. Sheep require much less, 2–6 kg of water per day on dry range, $0\cdot13$–$2\cdot7$ kg when being fed on rations of hay, roots and grain, and very little when grazing good pasture.[2]

Regular and frequent watering is essential for cattle. When water is rationed severely, milk production falls, and considerable expense in capital equipment to provide high-yielding cows with continuous supplies of water has been found justifiable. Sheep, in certain conditions, will also respond to good supplies of water. At the Desert Range Station in Utah it was found that sheep on the range gained an average of $1\cdot5$ kg each in a 40-day period when they were watered daily. They lost $2\cdot7$ kg each when they were watered only every third day.[3]

Water needs of animals vary, of course, according to their environment. The moisture losses from the body by evaporation increase with temperature, enabling the animal better to withstand high temperatures by getting rid of heat, but necessitating replacement of the water used up.

It is obvious why livestock industries flourish most easily in humid temperate regions, and in particular why dairying is severely limited in hot, dry regions. The adaptability of sheep is

[1] U.S.A. Department of Agriculture (1955), 359.
[2] U.S.A. Department of Agriculture (1955), 17.
[3] U.S.A. Department of Agriculture (1955), 15.

well known, but the difficulties of an environment like the desert margins of Australia will always clearly impose limitations on the productivity even of sheep, and necessitate adherence to a breed like the Merino, which is hardy, and produces a useful product — its fine fleece — in such conditions.

It is difficult to overstress the importance of water needs of plants and livestock, or the need to consider evaporation at the same time. Studies of the water balance concept recognise the function of storage capacity of the soil as a kind of bank, from which, directly and through the needs of plants, water is removed, and in which the funds are restored by precipitation. Since evaporation is related to temperature it is possible to use records of temperatures to calculate evaporation, and these combined with records of precipitation will yield a budget of water availability which can be maintained from day to day and month to month.

The classification developed by C. W. Thornthwaite[1] has been widely adopted for such practical purposes as calculation of amount of water to be applied in irrigation schemes, as well as being academically valuable for regional studies.[2]

Seasonal distribution of precipitation is hardly less important than the total amount. Plants need water most during their growing season, and hence, except to the extent that they can draw on water stored in the soil or receive artificial irrigation, it is during the growing period that rainfall is most needed. In countries where rainfall is seasonal, as in tropical monsoon and Mediterranean types of climate, late arrival of the rains may have serious consequences for food supplies. Failure of the season to produce normal rainfall, or early cessation of the rains, may similarly be serious.

The spatial as well as the temporal variation of the water budget has been intensively studied in the U.S.S.R. by Budyko and other workers.[3] Soviet atlases include maps showing the zonal pattern of the water budget and other agroclimatological data, emphasising the importance placed in the U.S.S.R. on the application of climatology to agriculture.[4] Even in the maritime climate of

[1] Thornthwaite (1948).
[2] Curry (1962) illustrates the value of Thornthwaite's formula in a study of pasture requirements and output in New Zealand.
[3] Budyko (1956).
[4] Abstracts of these data and simplified maps are given in Symons (1972).

Britain the need for irrigation has been shown to be widespread.[1]

Regularity of rainfall is also important for some crops. Thus for rubber trees (*Hevea brasiliensis*) the optimum rainfall is 1800–3800 mm, distributed so that no month has less than 75 mm. Tapping is hindered, however, on wet days, so that it is preferable that rainfall should be concentrated on not more than about 150 rain-days per year.

Destructive Aspects of Rainfall

So far we have discussed water only in its beneficial and productive aspects, but its destructive roles are also of importance in agricultural geography. In particular, rainfall is an agent of soil erosion. Geological erosion is a natural phenomenon and is beneficial in its exposure and transport of mineral elements and other materials which build up rock and soil when deposited. When we speak of soil erosion, however, we mean accelerated erosion, the removal of soil at a rate faster than it accumulates, so that the product of centuries or even millennia is dispersed in the form of non-productive and even destructive sediments.

Water, as it falls on to the surface of the land and as it runs off, erodes by splash erosion and scour erosion. Rain consists of drops of many different sizes, and heavy rain contains many of the larger drops, 3 to 6 mm in diameter. With high velocities, these have great erosive power through their kinetic energy when they strike the ground, and this may be increased by a driving wind.[2] The amount of soil set in motion by each drop is directly proportional to the square of its velocity. Particles of fine sand are moved readily even by fine rain, while heavy rainfall can lift particles of several millimetres. Rainfall intensity, soil texture, condition of the soil, and slope are the more important factors affecting erosive power. Similarly, erosion by run-off varies with the energy of moving water. The velocity of run-off on slight slopes is small compared with that of falling rain, but increases rapidly as slope becomes more marked, leading to rill, gully and sheet erosion. Protection against erosion calls for maintenance of vegetative cover, contour ploughing and construction of contour ridges, banks, stream control, etc. Erosion is most serious in climatic conditions which allow soil to dry out well between falls of

[1] Hogg (1967, 1970).
[2] Russell (1973) includes a summary of soil erosion and conservation.

rain, so making individual particles easily detached, and in such conditions it is normal for plants to be slow to colonise bared ground. The risk of erosion should inhibit many farming practices and ought indeed, in some regions, to be a major factor in deciding on the type of farming to be practised, but all too often warning signs are ignored and erosion is allowed to proceed until removal of soil is far advanced.

When water cannot drain sufficiently quickly off the land by percolation and run-off, flooding results. Even in areas of low rainfall and high evaporation, sudden rains of high intensity or prolonged duration cause severe flooding. In the arid regions of Australia, almost continuous drought is occasionally interrupted by catastrophic flooding. Erosion of topsoil and destruction of vegetation intensify flooding, by reducing infiltration of water into the soil and so increasing run-off. Protective works may have to be elaborate and therefore expensive. Long-term control usually resides in improvement of management in the watersheds and higher reaches of rivers, but these are precisely the areas to which it is difficult to direct investment, because such lands are in themselves generally of low value.

The characteristics of soils developed in waterlogged conditions and the formation of peat are discussed later, and there is further reference to accelerated erosion at the end of this chapter.

Snow

Where snow falls only for short periods each winter, and snow cover is intermittent, as in the British Isles, its main impact is in the hardship it places on livestock. Hill sheep farming in the Scottish Highlands, the English fells and the Welsh mountains is adapted to snow conditions and the sheep are moved to the safer areas when snow threatens. Even so, shepherds are frequently unprepared for sudden storms. Sheep can live for two weeks or so in snow drifts but rescue operations are arduous and losses in bad winters are heavy. On lowland farms also snowy winters may cause severe losses, and always lead to shortages of fodder. Heavy snow late in the winter is particularly serious as it may cause the death of large numbers of newborn lambs. Further losses occur when large masses of snow melt and floods result.

In countries where snow cover of longer duration is to be

expected, adaptation is more complete. In the Alps and the Scandinavian mountains, all livestock are housed, having been brought in from the summer grazings before the autumnal snowfalls make them unusable.

Snow insulates the ground from low air temperatures and so retards deep penetration of frost. In moderately cold conditions in the U.S.A. a depth of 30 to 45 cm of snow has prevented frost penetration and in severe freezing penetration has been avoided with a covering of 60 cm of snow. Soil freezing will often start before the accumulation of snow but if this builds up to a cover of 75 to 100 cm, and remains as deep, thawing will begin from the lower edge of the frozen zone as heat rises from below.[1] This will make the soil available more quickly for cultivation when the snow thaws. This is particularly important where spring-sown crops are grown. Land cannot be prepared for sowing until the snow has cleared and excess moisture dispersed.

Alternatively, advantage may be taken of the protection afforded by snow cover for suitable grains to be sown in autumn. However, although the snow protects the crop from frost and from drying winds, which when the ground is frozen may cause death through evaporation when there is no replacement water available, there are other hazards. Certain parasitic fungi find suitable conditions for breeding under snow cover and attack plant seeds. This is one of the factors discouraging cultivation of winter wheat and rye in northern Sweden.[2] On balance, snow cover is valuable in cold regions, and in the U.S.S.R. great attention is given to snow conservation. A winter with little snow is rarely followed by good harvests, particularly if winter wheat in the Ukraine is destroyed by frosts and the land has to be replanted by lower yielding spring grains.

Temperature

The germination of seeds and the growth of plants require suitable temperature conditions. The optimum temperature is usually between 18 deg. and 25 deg. C. Low temperatures permit only slow growth, the minimum for rye being about +2 deg., for wheat, barley and beet about 5 deg. and for maize about 9 deg. C.[3]

[1] U.S.A. Department of Agriculture (1955), 183.
[2] Fullerton (1954).
[3] Watson and More (1962), 32.

Since some plants need more warmth than others in order to mature, it seems that the amount by which each day's temperature exceeds the threshold should be significant for the plants grown and the degree of growth that takes place. Day by day throughout the growing season more day-degrees above the threshold will be added, so leading to the concept of accumulated temperature. Wheat, for which the threshold temperature is 5 deg. C. (41 deg. F.), needs about 1400 day-degrees of accumulated temperature. The marginal character of most of Scotland and Northern Ireland for wheat growing, based on this figure, is shown by maps of accumulated temperatures compiled by Gregory for the British Isles.[1] Rice needs about 3000 day-degrees and an average temperature during the growing period of at least 20 deg. C.

An example of a practical application of accumulated temperatures is their use in calculating the heating requirements for raising crops in glasshouses and apportioning costs between different crops.[2]

Maps of accumulated temperature should be of value in defining areas suitable for growing particular crops, in conjunction with field experiments. A limitation is the lack of suitable climatic statistics in many areas. However, whereas the tables for the calculation of accumulated temperatures prepared by the Meteorological Office[3] were designed to be used in conjunction with daily maximum and minimum temperatures, Gregory has achieved promising results with monthly means of temperatures, which are more widely available. Budyko makes extensive use of accumulated temperatures in his zonal maps of the U.S.S.R., combining them with the water balance to derive agroclimatological regions.[4]

Frost

In all regions where frost occurs it is one of the main enemies of crop cultivation and causes serious losses even in relatively mild areas such as the south of England. In more severe regions it dominates agricultural planning. The breeding of varieties of cereals which can mature in shorter frost-free periods is a major part of the battle to push cultivation further north in northern

[1] Gregory (1954).   [3] Meteorological Office M.O. Form 3300 (1928).
[2] Hogg (1970).   [4] Budyko (1956).

Europe, Siberia and Canada. The zone of permafrost, where soils are permanently frozen only a little below the surface, is similarly a zone of challenge in economic utilisation. Through Siberia permafrost extends as far south as Outer Mongolia, limiting the tayga forest to larch and other trees which can survive with a shallow root system developed in the soil zone above the permafrost. This zone which thaws in summer is low in nutrients and drainage is hindered by the frozen subsoil, so that cultivation offers small rewards and is limited to the most favoured pockets of land.

Frost is one of the chief factors in altitudinal limitation of cultivation. Least troubled by frost are locations close to the sea, which benefit from the amelioration of the climate by the proximity of sheets of water, and by the movement of air on the coast. At the other extreme are bottom lands into which cold air drains. Higher slopes may remain free of frost in such circumstances, and hence the preference for sloping sites for orchards, since fruit is particularly vulnerable to frost. Methods of protection against frost include burning of paraffin under the trees or spraying with water throughout frosty periods. Either method, but particularly the former, is expensive and laborious.

Abnormally late frosts are particularly damaging since they catch a high proportion of plants at the seedling or early leafing stages when they are most vulnerable. Recovery, however, may take place if the frost is not too severe or repeated. Otherwise the farmer's only remedy is to re-seed with later maturing varieties, which is not always possible.

Frosts are never met in lowland equatorial and many tropical regions. For example, frosts are unknown in Malaysia. The lowest temperature ever recorded there is +2 deg. C. in the Cameron Highlands.

Light and Sunshine

Light is essential for the growth of plants, the process of forming carbohydrates which make up a large proportion of their bodies being initiated by the intake (through photosynthesis) of energy contained in light from the sun. Direct sunshine is not essential, but is desirable for rapid growth and the ripening of crops.

Length of day is a factor in plant growth which enables

vegetation to grow rapidly in zones of high latitude. During the short summer when the sky is never dark plants can complete the flowering and seed formation essential to reproduction of the species. More rapid photosynthesis to enable useful plants to exploit the long day is one of the aims of plant breeding. Day length in the tropics is always below the critical length for some plants such as the potato and hops,[1] but many sub-tropical plants require ten hours of darkness for their flowering. For some species, however, the hours of darkness and light are unimportant.

In maritime areas cloud cover often reduces the amount of light theoretically available. This is of greatest importance in cold and cool temperate regions, where ripening of crops and their harvesting in a dry state is frequently in doubt. It is, however, also relevant in tropical regions. Although Malaya grows a considerable amount of rice for internal consumption and has for some years been endeavouring to increase production so as to be able to reduce imports, it has too cloudy a climate for maximum yields of rice. Cloud cover and excessive rainfall also restrict double-cropping which might otherwise be widespread.

Wind

Winds are of importance in agriculture chiefly for the increased rates of evapo-transpiration and consequent increased water needs which they produce, and the physical damage they wreak when excessive. On the other hand, the wind may be put to work on the land to drive windmills for pumping water or for the generation of electricity.

In regions subject to strong winds, like the British Isles, cereal crops are frequently blown over or 'lodged' and strength of stalk is an important factor in choice of variety sown. Very strong winds sometimes thresh the crop while it stands leaving only a straw residue to be harvested. 'Killing' winds in many countries are associated with particular directions of origin. The *mistral* of the south of France, a cold, northerly wind funnelled by the Rhone valley, sometimes brings heavy losses to growers of olives, citrus and other fruits on the Mediterranean lowlands. Large windbreaks are planted to protect the orchards. In contrast, the *sirocco* from the Sahara is a hot drying wind feared on north

[1] Webster and Wilson (1966), 26.

African coasts. Similarly, in southern regions of the U.S.S.R. the *sukhovey* is associated with the northern edges of anti-cyclones. Resulting drought and rapid increase of evapotranspiration can cause widespread losses of grain and other crops.

Hot, dry winds have commonly derived their character from descent after passing over mountains, the best known example being the *föhn* of the Alps. Another is the *chinook* of the Rocky Mountains, the onset of which will speed the melting of snow. This can be beneficial, particularly if it permits early cultivation but if severe frosts come after the snow has gone autumn-sown seeds will no longer be protected.

Strong winds, and particularly those that are dry, are also important agents of soil erosion. The most publicised of all cases of soil erosion, the 'dust bowl' of Kansas and Oklahoma, was the work mainly of wind on bared soil. Movement of soil particles is by suspension, surface creep, and saltation — the particles leaping and bounding. Reduction of surface velocity is one of the main principles in control of wind erosion. The force of wind varies with the square of its velocity and so even a small reduction in velocity is valuable. This can be achieved through keeping the land covered, or making the soil up in banks and by planting shelterbelts.

Fallowing, as practised in dry farming, when the land may be left without a crop for 20 months or so in order to conserve moisture for the following crop, clearly lays the soil open to erosion. This is the more likely since the practice of dry farming implies that the soil will be so dry as not to clod readily. It is, of course, the finest soil particles that are most vulnerable, while small clods of soil are particularly valuable because they have a high protective surface in proportion to their weight.[1] To be resistant to erosion, they must be larger than a wheat kernel or bean, and to contain at least two-thirds by weight of non-erodable fractions. These conditions are most likely to be met in soils of medium texture, with clay content of 20 to 30 per cent and a high silt content.

Unless protection can be seen to be adequate, bare fallow, and, if necessary, cropping, should be avoided and the land kept in grass, which gives a good protective cover and is well anchored.

The ploughing of 20 million hectares of the Soviet Union's

[1] U.S.A. Department of Agriculture (1957), 312.

virgin lands between 1954 and 1965 led to serious soil erosion in
Kazakhstan and Siberia, which was only checked when balanced
rotations and livestock rearing were introduced on these lands.

SOILS

Since the soil is the essential material on which agriculture is
based, any comprehensive survey of the geography of agriculture
should include a fairly thorough treatment of soils. This is not
practicable within the scope of this book but there are many
excellent texts[1] to which the reader may refer, as indeed there are
for climate and other factors only briefly discussed. Therefore, in
this section as in others, treatment will be confined to discussion
of some points which seem to be particularly in need of stressing
in any approach to agricultural geography.

Soil is composed of four major constituents: inorganic particles, organic material, water and air. The factors, active and
passive, that determine the character of a soil may be grouped as:

(a) Parent materials   (d) Vegetation
(b) Climate   (e) Soil fauna and bacteria
(c) Relief   (f) Man's utilisation.

All of these act through time, which may also be considered a
soil-forming factor, though not separately considered here.

Parent Materials

The soil is formed from the rock on which it develops, together
with other materials which are deposited on the site in the course
of its formation. The latter include organic materials derived
mainly from the decay of plants, of which some of the more
primitive kinds can obtain a foothold on bare rock as soon as it
begins to weather, and so contribute to its erosion and the
formation of soil. Partially-decomposed organic matter which is
being incorporated in the soil is called 'humus', and it combines
chemically with some of the mineral products of the weathering
of the rock. Rock is weathered by mechanical and chemical
agents, the resulting partly weathered material being known as
the *weathering complex*. The character of the weathering complex

[1] Russell (1973) is particularly recommended for agricultural aspects of soils,
Bridges (1970), Selby (1971), Cruickshank (1973) and Curtis, Courtney and
Trudgill (1976) for geographical approaches. Cruickshank provides a chapter on
soil as a resource.

and of the soil that develops on it depends partly on the character of the parent rock. The great variety of soils found on differing geological strata in close proximity led to early soil scientists giving rather too great an emphasis to the role played by the underlying rock, to the neglect of the influences of climate and vegetation. Nevertheless, lithology does have great importance, and this is especially so from an agricultural point of view, because the texture and chemical content of a soil are largely dependent on the parent material, and these are obviously of fundamental importance to the farmer.

Texture varies with the proportion of clay, silt and sand particles which make up the mineral basis of the soil. These particles are classified by the U.S. Department of Agriculture as follows:

	Millimetres
Clay	Below 0·002
Silt	0·002–0·05
Very fine sand	0·05 –0·1
Fine sand	0·1 –0·25
Medium sand	0·25 –0·5
Coarse sand	0·5 –1·0
Very coarse sand	1·0 –2·0

Soils are classified according to their different combinations of sand, silt and clay by mechanical analysis, grouping being shown graphically in texture-triangles (Figure 1).

Accumulations of materials of the extremes in texture — pure sand, silt or clay, cannot be regarded as soils, but they may in the course of time develop into soils with the aid of the various agents, in which case, of course, the texture changes. Sandy soils are generally infertile because they are incapable of retaining needed chemicals in the soil. Sandy loams, however, with their higher percentages of clay and better structure, may be very valuable soils, especially in areas where soils which are 'heavy' i.e. containing much clay, are liable to waterlogging. They are easily 'worked' and warm up early in the spring. They do not, however, have the potentiality for high fertility that is contained in soils with more clay. Medium loams are often regarded as ideal soils because of their combined qualities of retentiveness of water without tendency to waterlogging, normal richness in plant

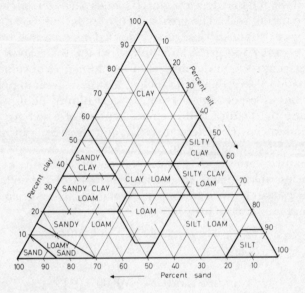

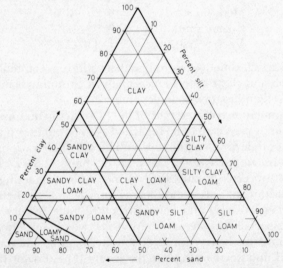

Figure 1. Texture triangles, used for the classification of soil according to mechanical analysis by the U.S.A. Department of Agriculture (above) and by the Soil Survey of England and Wales based on the particle-size grades of the British Standards Institution (below).

foods and receptiveness of fertilisers, and general ease of working. They also offer the maximum alternatives in switching from arable cultivation to grassland, and the most even and sustained productivity in a crop and grass rotation.

The greatest inherent fertility and suitability for some crops resides, however, in soils with rather high clay fractions. Provided drainage is adequate, or is made adequate by the insertion of drains, some 'heavy' soils provide not only highly productive and durable pastures, but also excellent land for growing wheat and other crops with a high level of demand for nutrients. This is because of the virtues of the clay particles. Whereas sand particles are chemically inert, clay particles are chemically active. Intricate chemical changes occur between them and the organic materials in the soil to form the *clay-humus complex*.

The clay fraction is made of clay minerals, such as kaolinite and montmorillonite, which are composed of sheets of alumina and silica bonded by oxygen atoms in respectively $1:1$ and $2:1$ ratios. Through ionic substitution of atoms, the clay crystal receives a negative charge. This results in attraction between the clay particles and other particles with a positive charge, e.g. hydrogen, calcium, magnesium, sodium and potassium ions. The attracted ions are held in the soil solution but in response to changes in this can be released from the soil particle or 'exchanged'. Such exchangeable positively charged mineral ions (cations) account for the *base status* of a soil.[1]

A clay-humus complex which is rich in exchangeable bases is potentially a fertile soil. It is, however, continuously exposed to the *leaching* action of rain water, which is a dilute solution of carbonic acid (H_2CO_3) able to exchange its hydrogen ions for the nutrient ions in the soil. These are then carried away as dilute salts in the drainage, and the clay-humus structure is increasingly left with hydrogen ions which make it acid. The desired nutrient ions must therefore be replaced by further mineral particles from erosion or organic decomposition. To remedy deficiencies caused by absence of, or slowness of, the natural supply, manure and chemical fertilisers are added.

The proportion of exchangeable bases in a soil is obtained by the process of measuring the concentration of hydrogen ions. It is assumed that the proportion of other ions which can be held by

[1] For a further but brief description, see Eyre (1968).

the clay-humus complex depends on the 'space' left by the hydrogen ions. The proportion of free hydrogen ions in the soil solution is measured and stated as the pH value.[1] pH7 is neutral, values below this indicate an acid soil, values above indicate alkalinity. Most agricultural soils are between pH5 and pH8. Some moorland and other extremely acid soils are of pH4 or $4 \cdot 5$.

The pH value gives an indication of the potential value of a soil. A low pH value, indicating an acid soil, will be sufficient to discourage planting on that soil of lime-loving crops unless lime is added, while a high value may indicate unsuitability for a plant that grows best under acid conditions. At the same time it must be realised that pH indicates only relative acidity. A soil with a high clay fraction and a relatively low pH may have more exchangeable bases than one with a low clay fraction and a high pH value. For a full picture of the soil it is therefore necessary to have measurements of each of the main exchangeable bases, or *cation-exchange capacity*.

Cation-exchange capacity of a soil, as already noted, will vary according to the texture of the soil. Sandy soils will have low cation-exchange capacities because the sand particles provide only a skeletal structure and the relatively small amounts of humus and clay present will not be able to offset this inherent weakness. Silt particles, it should be added, are intermediate between sand and clay in this respect, but nearer to the sand fragments in character, with very limited chemical activity. A higher cation-exchange capacity will be found in a clay soil, but this will vary according to the type of clay minerals present. Thus, montmorillonoid materials have large cation-exchange capacities, whereas those of kaolin minerals are small. Soils high in organic content have high base-exchange capacities because the humus particles develop the required large negative charges.

The cation-exchange capacity is usually expressed as milli-equivalents of cations required to neutralise the negative charge of 100 grams of soil with the pH held at 7. An example of this kind of analysis appears in Table 1.

The exchangeable cations vary from one soil to another, according to natural characteristics and past management. Sodium is found in high proportions in strongly alkaline soils,

---

[1] Russell (1973) gives details of pH measurement and its difficulties, 121–128.

calcium and magnesium in nearly neutral soils, and hydrogen and aluminium in acid soils.

From the foregoing it will be evident that the parent materials contribute greatly to the character of resulting soils. Since most sedimentary rocks are the product of materials previously sorted by suspension in water or air, it may be expected that the weathering complex derived from a sedimentary rock will be less varied in mineral content than one derived from an igneous or metamorphic rock. Sandstones will yield soils with a high proportion of quartz particles, so that the texture will be sandy,

TABLE 1

CHEMICAL ANALYSES OF TWO NEW ZEALAND SOILS

Soil Locality and grid reference	Paparua sandy loam 2 miles WNW of Winchester: S102. 772787		Templeton silt loam 3 miles N.W. of Winchester: S102. 768808	
Depth (cm)	0–15	35–50	0–15	35–46
Horizon	A	B	A	B
pH	6·0	6·5	6·1	6·3
1% citric acid soluble P_2O_5%	0·035	0·037	0·020	0·007
Organic C %	3·4	0·6	2·8	0·55
Total N %	0·29	0·09	0·26	0·08
$\frac{C}{N}$	12	7	11	7
Exchange capacity m.e. %	17·1	13·0	14·8	11·2
Total bases m.e. %	11·7	11·2	11·2	9·0
Base satn. %	68	86	76	80
Ca. m.e. %	9·3	10·3	8·7	7·2
Mg. m.e. %	2·3	1·3	2·0	1·7
K m.e. %	1·00	0·30	1·50	0·35

Source: Soils and agriculture of part Geraldine County New Zealand, Soil Bureau Bulletin 13, New Zealand Department of Scientific and Industrial Research, 1959, 18.

whereas shales tend to break down into clays. Among igneous rocks, granite breaks down into sandy soils which are generally poor in nutrient minerals, whereas basalt, a basic rock, will tend to yield clayey and potentially fertile soils.

Climatic Influences on Soil Development

Although certain characteristics of fundamental importance in the soil depend on the parent material, just how these characteristics will be developed depends on other forces, of which climate is the one which is universally important. During the last two decades of the nineteenth century, V. V. Dokuchayev and other

Russian soil scientists recognised the dominant influence of climate through the persistence across the great spaces of Europe and Asia of latitudinally arranged soil belts. The concept became generally accepted with the discovery that similar relationships between soil and climate could be seen in North America and elsewhere on the continental scale.

Soils that thus correspond with the great climatic belts are called *zonal* soils. Those that differ because they are derived from specific parent materials, such as limestone, or because of the presence of salt or excess water in the profile are called *intrazonal*, while those in which time has not been sufficient for recently deposited materials to weather in the form appropriate to the climatic zone are called *azonal*.

However, after dominating soil classification for over sixty years, the zonal concept is no longer in general use in soil survey at the national level, though of value in considering soils on the world scale. It has been largely superseded by classification based on the study of carefully defined diagnostic horizons and other features of the soil profile, as developed principally by the United States Department of Agriculture. National soil survey organisations in a number of countries have been influenced by the American Soil Survey as their new classification has been evolved, and with the publication of *Soil Taxonomy*[1] it is likely that it will have increasing effect. Many countries retain their own systems of classification, however, and even where some of the American concepts have been adopted much of the traditional terminology has been retained.[2]

The need to facilitate use of soil survey findings in agricultural and forestry work favours retention of familiar terms and concepts, and there is proven value in basing description on the profile. The practical significance of the profile may be illustrated by the podzol (Figure 2) and the variation that occurs in podzolic soils according to the detail of climate, relief, vegetation and usage. Podzols are subjected to persistent leaching, so that the

[1] U.S.A. Dept. of Agriculture Publn. No. 436 (1975). This is so also in the FAO/UNESCO Soil Map of the World now in course of publication at a scale of 1:5,000,000.

[2] Avery (1973) described the modified system as used by the Soil Survey of England and Wales. For briefer descriptions and classification comparisons see books noted on page 32.

bases are carried down from the eluvial A horizon (topsoil), characteristically ash-grey in colour, to be re-deposited in part at greater depth in the B horizon of illuviation (subsoil). Such a soil is not naturally fertile. Furthermore, the parent material of much

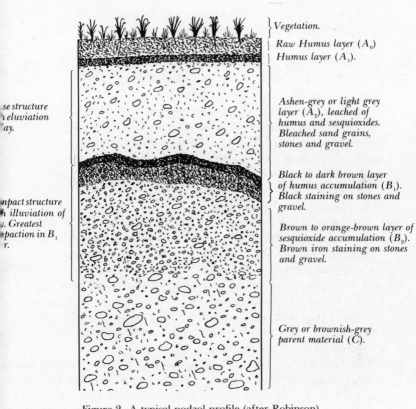

se structure
eluviation
ay.

npact structure
illuviation of
. Greatest
paction in B₁
r.

} *Vegetation.*

} *Raw Humus layer (A₀)*
} *Humus layer (A₁).*

} *Ashen-grey or light grey layer (A₂), leached of humus and sesquioxides. Bleached sand grains, stones and gravel.*

} *Black to dark brown layer of humus accumulation (B₁). Black staining on stones and gravel.*

Brown to orange-brown layer of sesquioxide accumulation (B₂). Brown iron staining on stones and gravel.

Grey or brownish-grey parent material (C).

Figure 2. A typical podzol profile (after Robinson).
Recent practice would prefer the symbols L (litter), F (fermentation), H (humus), Ea (bleached horizon), Bh (enriched with humus), Bs (with sesquioxide).

of the podzol zone is glacially-smoothed rock which has not had time since the retreat of the ice for the weathering that would release fresh nutrients, while the natural vegetation that has contributed to the growth of the soil is usually coniferous forest or scrub with acid leaf-litter. To be agriculturally productive,

podzols need the addition of lime, and commonly other nutrients, including phosphate, potash and nitrogen. When a podzol
is ploughed and manure or 'artificial' fertilisers have been added
so that the soil becomes fertile and the horizons become
intermixed, the soil may be described as an 'agricultural brown
earth'.

The brown earths, or brown forest soils, which are the product
of less intensive leaching and a covering of deciduous forests, are
by their nature more promising agricultural material. Brown
earths which are also loamy in texture are perhaps the most
productive of the world's mature soils under the systems of mixed
farming that they are able to support.

On the other hand, whether brown earth or podzol, a soil in a
cool and humid climate in natural or cultivated conditions may
develop in its illuvial horizon a layer of impermeable material —
hard pan or iron pan — which impedes drainage. Unless this is
broken up, the soil will eventually become waterlogged, especially
if it is not covered by trees, which are efficient at intercepting rain
and consume much soil water.

Land which is saturated for long periods has gleyed soils —
characteristic green and blue mottling distinguishing the profile.
In cool and humid climates, such as in north-west Europe, gleyed
soils are also impoverished soils, leached of mineral bases, and
acid — pH value $4 \cdot 0$ to $4 \cdot 5$ compared with the neutral value of
$7 \cdot 0$. They are accordingly low in productivity, and permanent
pasture, often with rich but unwanted growth of rushes, is the
characteristic land use. Improvement can be effected only after
provision of adequate drainage.

Where a high P/E ratio allows water to remain long on the
surface, drainage improvements must extend from the individual field to the outfall of main watercourses at the sea. Thus,
whereas if soil erosion is an acute danger ploughing should
always follow the contour, where soils are constantly moist and
the main need is to drain off surplus water it is perfectly legitimate
to plough up and down slope. Field ditches, minor streams, rivers
and canals must be dealt with if increased run-off at one stage is
not to result in worsened flooding or saturation at lower levels.
Pumping may be necessary to speed the flow.

More or less permanent saturation leads to the growth of
vegetation which is anaerobic, i.e. tolerant of the absence of

oxygen. Fen, bog and swamp are some of these conditions, the former two being distinguished by the formation of peat from partially decomposed vegetation.

Peat, which is not, strictly speaking, soil, is of two main kinds, acid and basic. Acid peat occurs in regions of cool and humid climates and may cover large tracts of hill and mountain country, being then known as blanket bog. Drainage may not be difficult, owing to the relief of the land, but harsh climatic factors and acidity (pH $3 \cdot 0$–$4 \cdot 5$) of the peat soils make the land of low agricultural value. Forestry is an alternative use for improved blanket bog in the British Isles. Basic or fen peat may accumulate in depressions into which water drains, if this water contains sufficient lime. This peat has high potential value and when drained may produce some of the highest class arable and horticultural land, as in the fenlands of eastern England. Reclaimed peats also provide valuable soils in very different climates, as in Malaysia, where they are used for growing pineapples.

For peat to form under tropical conditons, where evaporation is high, not only must drainage be poor but rainfall must be very heavy and well distributed throughout the year. Under heavy rainfall and with good drainage in tropical conditions, the podzol of the cool regions has its counterpart in tropical red and yellow podzolic soils. The red and yellow colours are believed to be caused by the presence of sesquioxides, left when silica is leached out. In tropical conditions, however, the horizons associated with podzolisation are commonly absent, and there is a great depth of a comparatively uniform accumulation of fine clays, rich in aluminium and iron oxides and low in silica. These red earths may have been derived by a different process known as ferrallitisation, and the soils are often called 'ferrallitic', the term 'lateritic' having been abandoned owing to the confusion which had developed in its use. The name 'laterite' is still used for the material in some of these soils which hardens irreversibly on exposure to the air, severely limiting agriculture. The depth and colour of the tropical red earths and the luxuriant vegetation which grows rapidly upon them led early investigators into the error of believing that such soils were inherently fertile. In fact the excessive rainfall rapidly leaches out the deposited plant nutrients, resulting in low fertility. The more fertile tropical soils

are the young, immature, alluvial soils which are almost devoid
of profile changes, and in which renewal of nutrients from
river deposition is regular and frequent. Good examples
of the periodic flooding of alluvial plains by silt-bearing waters
occur in the basins of the Mekong and Chao Phraya in south-
east Asia.

Relief as a Factor in Soil Development

Mention has already been made of relief as one of the factors
influencing the amount of water that infiltrates the soil. This
relationship may be seen by comparing a flat and a rolling area.

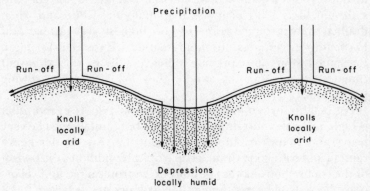

Figure 3. Variation of supply of water to the soil with relief.
Source: J. H. Ellis, *The soils of Manitoba.*

The flat land receives whatever precipitation falls on the area, but
water reaching the rolling area varies between the knolls which
receive the precipitation less run-off and the hollows which
receive precipitation plus run-off. The soils of the knolls are
locally arid while those of the depressions are locally humid
(Figure 3).

The extent to which moisture is retained in soil on a slope
depends, of course, partly on the texture of the soil. Soils with a
very high clay content in a cool, humid region can become
waterlogged even on quite steep slopes. In Ireland, on slopes of
10 degrees on drumlins, marked gleying of soils is found. The
water table is at greater depth high on the slope than lower down,
and between the drumlins it may be at the surface, creating marsh
or bog.

Soil on a slope is subject to the forces of gravity like any other mass and consequently even in conditions of great stability there is some movement downslope of soil particles. Movement is initiated mainly by running water, frost action, wind, falling vegetation or the passage of animals. It is greatest where slopes are bare, least where they are well vegetated. When natural conditions of comparative stability are upset, accelerated erosion may become serious, taking the form of sheet, slip, rill and gully erosion of the mineral soil. In severe conditions the whole of the topsoil may be stripped off, with erosion continuing into the subsoil.

Downslope movement will always result in soil at the bottom of a slope being deeper and richer in nutrients than higher up — unless of course it is being scoured away at the foot. Cultivation increases movement and, therefore, the contrasts in depth of soil. This is particularly so if ploughing is carried out up and downslope, as is common in humid regions, where even sloping lands are likely to remain excessively moist for long periods.

Vegetation and Soil Development

As noted above, the role of vegetation in the development of a soil has been underestimated in the past. In fact, in natural conditions soil and vegetation evolve in intimate relationship. Thus, with the same climate and parent materials, a difference in grazing pressure will result ultimately in different vegetation. For example, a fence protecting woodland from grazing may ensure survival of the trees, while on an adjacent area regeneration is prevented by grazing and pasture develops, unless drainage is poor and the land becomes marshy. In each case the course of further soil development will be different, according as there is different leaf-fall, infiltration and evaporation of water, root development and other factors. After a long period of time the soils of the three land-use types — woodland, pasture and bog — become respectively podzols, gleys and peats. If agricultural development is then applied to all three the techniques required will vary and the potential productivity will vary, at least for a considerable time.

Soil Fauna and Bacteria

Animals that live on, or in, the soil necessarily affect its development in many ways directly as well as through their effect on vegetation. If they are burrowing animals they open up channels under the surface, which, in moderation, may be beneficial in facilitating rock weathering and soil aeration, but in excess may contribute to accelerated erosion. The work of earthworms is of great importance, their burrows extending often several feet into the ground and the passing of soil through their bodies having an important effect on its structure. The addition of earthworms to areas deficient in them has become recognised as a means of improving fertility.

Bacteria occur in their millions in soils and carry on beneficial work. They contribute to the decay of plants and the 'processing' of the remains of animals and their waste products. They liberate carbon dioxide which promotes rock weathering and some kinds of bacteria fix nitrogen. Although plants are surrounded by nitrogen in the atmosphere they are unable to use it directly. Their essential needs can only be met through the soil and thus bacteria that form suitable compounds (nitrates) are most beneficial. The nitrogen-fixing value of clovers, beans and peas is well known, these plants having nodules on their roots which contain colonies of nitrogen-fixing bacteria.[1]

Different soil bacteria are found in different conditions. For example, in anaerobic conditions there will not be the same bacterial life as in well-aerated soils, but other forms will occur, especially those which live in the humus characteristic of gley and peat soils. One of the functions of adding lime to soil is encouragement of desirable types of bacteria and the work they perform on soil structure and chemical content.

Effects of Man's Utilisation on Soils

Relationships between a soil and the soil-forming factors being as intricate as they are, it will be apparent that the introduction of man on the scene, especially modern man with a great range of crops, livestock, machinery and chemical compounds at his disposal, will produce far-reaching changes. In extreme cases, such as the conversion by prolonged cultivation and fertilisation of comparatively poor, sandy soils into market-gardening land, it may be difficult to ascertain the original nature of the soil. Nevertheless, many such developments are impermanent and if

---

[1] Russell (1973) gives a comprehensive description with illustrations, 357 ff.

attention is relaxed reversion to lower fertility occurs. Attempts to change the character of a soil are sometimes hazardous and always expensive, and so are limited in practical farming. Although great changes are eventually made in soil even in normal cultivation, the soil on which work is commenced in any season is basically the same as it was the previous season. The relatively permanent characteristics of soils make possible classification which is of practical value to farmers.

For practical purposes the classification of soils mapped in field survey is made on the basis of qualities of the individual profile. Soils with similar profiles are grouped together as *soil series*. These in turn are grouped together into higher orders — in England and Wales into *Subgroup, Group* and *Major Group*. The higher group will indicate certain features of agricultural importance, which may include common nutrient deficiencies and response to lime, fertilisers and trace elements, but thorough appreciation of the potential of land (discussed in Chapter 10) demands investigation of the soil series.

A soil series is distinguished by a name made up of a place name, perhaps the place where it was first identified, or is typified, and the soils included in the series will have similar horizons developed from lithologically similar parent materials and will be similar in texture, depth and drainage. The soil series will therefore be found only in similar physiographic conditions. It may be subdivided, where one characteristic varies, into *phases*, e.g. stony phase, very stony phase.

In England and Wales the soil series name is also used to describe the *soil mapping unit*, an area within which the majority of soil profiles are those of the dominant series. A group of topographically related soils developed on one geological parent material may be called a *soil association*, a term used mainly in mapping Scottish soils. Terminology does vary from country to country and is evolving continuously; so interpretation for the farmer, forester and planner requires considerable attention.

In agriculture, man is continuously faced with the problems posed by his physical environment. Whatever the resources of factory and laboratory at his disposal directly or indirectly, he must always apply these resources according to the nature of the land on which he works and the climate in which he works. He will, however, be able to sustain and increase the yield of the land if the social and economic environment permits him to take advantage of the opportunities that are offered by the physical environment as interpreted in the light of technological innovation.

RELIEF

The relief of a land surface affects agricultural utilisation through (1) altitude, and (2) slope. The effects of altitude are felt mainly indirectly, i.e. through climate, while slope controls are partly indirect through climate and soil, and partly direct, such as limitation of cultivation by steepness.

Climatic Effects of Altitude

The primary consequence of high altitude is lowered air pressure, because of the reduced amount of the atmospheric 'envelope' above a raised surface. Only in exceptional cases does this have direct agricultural significance, because the effects of decreased pressure are generally evident only above the levels at which climatic controls limit land utilisation. The secondary effects of decreased mean temperatures and increased precipitation and wind forces are the economically important consequences of higher elevations.

Mean temperatures decrease with altitude because, with the lower content in the rarified air of carbon dioxide, moisture and other particles, the sun's rays pass through it with less warming effect. Consequently, a high proportion of solar radiation reaches the ground surface which heats up rapidly in the sun. This in turn may cause extreme desiccation. Conversely, the ground loses heat rapidly in the free radiation of night. The effects of increased altitude are thus not directly similar to those of high latitudes, where the sun's rays, because of their low angle to the surface of the ground, have always to penetrace a thicker 'blanket' of atmosphere and be more diffused over the surface. On the other hand, high latitudes derive some compensation from long days in their summers. The net result is that there is some similarity in the life forms of natural vegetation in their zoning respectively through latitudinal and altitudinal belts. At both sea-level in cold temperate regions and 4500 to 6000 metres in some equatorial regions, for example, needle-leaved coniferous trees are dominant forms of vegetation if other environmental conditions permit Similarly, within these latitudinal and altitudinal zones the hardiest cereals find their limits and livestock herding forms an important source of food and raw materials.

At intermediate latitudes the altitudinal variation is correspondingly adjusted. In the Himalayan ranges — outside the tropics — wheat and barley are cultivated above 3000 metres while summer pasturage is found at 3500 to 4500 metres. In the French and Swiss Alps the summer grazings are usually between 1800 and 3000 metres. 1800 metres is roughly the limit of tussock grasslands in New Zealand at a latitude of about 44 deg. S. In the British Isles, mountain sheep find grazing to about 1000 metres, and coniferous forests find their limits at about 600 metres in the east and much lower in the west where wind is an important factor inhibiting growth. Hay, oats and potatoes are grown up to about 300 metres in favourable circumstances. In the south of Scotland the growing season decreases from about 8 months at sea level to 6 months or less at 300 metres and $4\frac{1}{2}$ months at 600 metres.[1]

The handicap imposed on crop growing by higher altitudes, and high latitudes, is not simply one of restricted vegetative growth but of difficulties with ripening. In the Alps, retarding of the harvest has been recognised as a day for every 30–40 metres of altitude.[2] This emphasises the value of crops which are useful for stockfeeding even when not fully ripened, such as oats and barley.

The decrease in temperature with altitude is not regular but depends on many complex factors. The actual lapse rates vary widely, and temperatures on the ground surface will vary still more widely according to conditions of slope, aspect and exposure. In the British Isles the figure of 6·5 deg. C. per 1000 metres or 1 deg. F per 300 feet is accepted as a reasonable average, but three times this figure has been recorded on British mountains between sheltered valleys and exposed sites. The lack of high level stations recording temperature data has hindered investigations into correlations between the altitude limits of different crops, dates of sowing and harvesting and other aspects of altitude. Some tentative conclusions for Northern Ireland are reproduced in Figure 4. In this diagram the values of accumulated temperatures for selected lowland stations have been plotted (Graph A). Similar values were computed and plotted for a series of heights above the meteorological station at Aldergrove

[1] Halstead (1958). [2] Peattie (1936), 25.

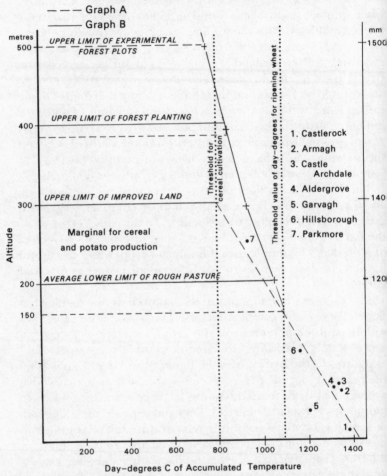

Figure 4. The relationships between altitude, climate and the use of land for crops. An example from the British Isles.

Based on a diagram by N. Stephens in *Land Use in Northern Ireland* (ed. L. Symons).

(66 m) based on the above lapse rate. The diagram suggests that the upper limit for cereal production (in terms of day-degrees of accumulated temperatures) is here reached at about 300 metres. In fact, only isolated fields of improved land are found as high as this and the normal limit of cultivation is about 200 metres. When famine conditions prevailed in the nineteenth century, crops were grown at higher levels than is regarded as economic today, but the present low altitudinal limits suggest that conditions are more severe than the graphs indicate, contributing factors being high rainfall, humidity and wind speeds.[1]

The increases in rainfall and wind with altitude may be no less important for agriculture than are decreased temperatures. Since the capacity of air to hold moisture varies directly with its temperature, and this decreases adiabatically in ascending air, high land is commonly subjected to heavy precipitation. This is particularly so, of course, where prevailing winds carry a great deal of moisture and are very persistent, as in the westerlies in the temperate zones. This is well illustrated in the South Island of New Zealand. Annual rainfall on the westerly slopes of the main divide (2500 to 3700 metres) varies between about 2000 and 10,000 mm. On the foothills of the eastern slopes it is only about 1750 mm and this falls to 600 mm or so on the Canterbury Plains. On the wet, western slopes, temperate rainforest is dense, and when this is cleared agriculture is hampered by the rainfall, while former forest on the eastern side has given way entirely to arable and grassland farming.

The combination of sufficient altitude and moisture results in snow, which severely impedes agriculture in most mountain areas in temperate regions. Mention has already been made of measures adopted in the European mountains to meet this problem.

Precise information on windiness at high altitudes is limited partly because of the high cost of the instruments necessary for recording. Further, anemometers are usually located to serve the needs of meteorological research, engineering or other occupations which do not reflect conditions on agricultural and pastoral land. A British survey related to the generation of electricity by wind power however, obtained records from summits at between 300 and 700 metres which rose comparatively little above

[1] Stephens (1963).

surrounding hill and moorland grazings and were, in fact, also grazed. The results showed averages of between 35 and 40 km/hr throughout the year.[1]

High winds act as a further deterrent to the growing of crops on high land and also adversely affect livestock rearing. To withstand the harsh conditions, cattle and sheep grazed on the hills must be of the hardier breeds which are usually slower to mature and less productive than lowland breeds. Moreover, the danger from snow is much increased by drifting.

To reduce the effects of high winds, shelterbelts are desirable in exposed areas, but it is in just such areas and partly because of the wind that trees are difficult to establish, and liable to be destroyed by exceptional gales just when becoming useful.

Mountain Soils and Vegetation

As in other conditions, the soil profile summarises the various environmental conditions acting over time. The soils of mountain and hill regions in the temperate and cold zones are poorer in nutrients than those of lower areas. In a region where forest brown-earths are the soils of the lowlands, hills of height sufficient for the climatic deterioration to be marked will carry podzolised or skeletal soils. In regions where the lowland soils are podzols, as in the north of the British Isles, profiles on the hills commonly degenerate into peaty podzols and are ultimately replaced by blanket bog. Acidity characterises both mineral soil and peat, a condition reflected in the calcifuge vegetation, dominated by species such as heath (*Calluna vulgaris* and *Erica spp.*), rushes and sedges (*Juncus, Carex* and *Scirpus spp.*), coarse grasses, such as purple moor-grass (*Molinia*) and the moor mat grass (*Nardus*), and mosses, notably the bog moss (*Sphagnum spp.*). Few of the species are agriculturally of much value, though heather plays an important part in the feed of hill sheep, especially when more nutritious grass is lacking or covered by snow.

The possibilities of improvement through grazing control, leading to better pasture utilisation and improved animal nutrition, have become more fully understood in recent years. It has been suggested that where a hill farm has 15–25 per cent of its area in bent-fescue grassland it could be made more productive

[1] Tagg (1957).

simply by enclosing some of this land to effect a change in pasture composition through controlled grazing. With *Nardus-Molinia* grass-heath, improvement can also be achieved in this way but will be greatly accelerated by improvements in the soil base-status.[1] On peat-covered areas there is little alternative to a substantial degree of soil improvement followed by reseeding, though the extreme acidity and the rapidity with which lime is washed out of the soil militates against the conversion of acid moorlands of this type into improved grassland. British hill pastures normally require the addition of five tonnes of ground limestone per hectare before any substantial improvement can be effected, and this is rendered difficult and costly by the remoteness of the grazings and the difficulty of access caused by steep slopes and bogs. In other conditions, as in New Zealand, where the addition of 300–400 kg of fertilizer per hectare together with seed, is normally a satisfactory top-dressing for the improvement of low-tussock grassland, the use of aircraft has overcome some of the problems created by relief.[2] So far aerial top-dressing has proved of little value in overcoming the lime-deficiency problems of the British moorlands and mountains but has been of great value in ridding the more fertile areas of bracken.

Increased altitude does not always mean poorer conditions for agriculture. In tropical conditions the reverse is commonly true, the temperatures and humidity of low lying and coastal regions being unfavourable to many crops and most kinds of livestock as well as human beings. Here it becomes a question of the needs of individual crops. Thus, in Java, where there is sufficient relief of land there are vertical zones of crops corresponding to the altitudinal zones of vegetation. Further differentiation takes place according to the variation of rainfall and other factors, but sugar cane is confined to low ground (below 200 m generally), rubber is found from sea-level to about 600 metres, while the highest yields and best qualities of tea are obtained from between 1000 and 1800 metres. But higher yet, on mountain ranges, the deterioration associated with montane and alpine zones occurs as in temperate countries.

Exceptions also occur in the details of soil pattern, associated with local variations in relief and parent materials. Well-drained, basic volcanic soils, for example, are more fertile than the soils of

[1] Edie and Cunningham (1971).   [2] James (1971). See also p. 128.

other parent materials which may lie below them at more promising altitudes.

Slope

As already noted, the effects of slope on agriculture may be considered as acting either directly or indirectly. With regard first to the indirect effects, some pedological and climatic effects of slope have already been considered, including the position of the water table, downslope movement of soil, air drainage and relative freedom of slopes from frost.

At any altitude the climates of different slopes vary according to their aspect. In the European Alps many studies have been made of the contrast between the sunny slope (French *adret*) and the shady slope (*ubac*). In such regions of high relief the contrast resulting from the angles at which the sun's rays strike the ground surface is heightened by the difference between the number of hours of possible sunshine that can be received by different slopes. Where cloud cover is usually of comparatively short duration, so that free radiation prevails during much of the time when the sun is not warming the land, the contrast is heightened. For plant growth it is the difference in soil temperatures rather than air temperatures between the *adret* and *ubac* and at different altitudes which appears to be significant.

Methods have been devised[1] to present cartographically the contrasts between the local climates of sunny and shady slopes and other features of insolation in mountainous terrain. Cartographic analysis is necessary to correct unsatisfactory generalisations which are too easily made regarding the value of the contrasting slopes. In the Defereggen Tal (Hohe Tauern massif), for example, shady slopes of gullies and embayments were found to be entirely used for forest and meadow while south-facing slopes, however small the area, were used for cereals. Calculations for intensity values in other localities in this region, however, generally show that cereals *can* be ripened when intensities fall below these gully values.

> This illustrates, therefore, what might be described as the *selective* rather than the *prohibitive* influences of insolation, for the economy of the valley demands a high meadow

[1] Garnett (1937).

acreage, and as there is often a better crop on a shady rather than a sunny slope, the concentration of cereals on slopes of one aspect and of meadow on another merely reflects the selection of slopes to which each is best — but not exclusively — adjusted, in terms of both insolation and the local economic requirements of the commune.[1]

In the siting of settlements and the cultivation of crops, long duration of sunshine was seen often to be more important than its intensity. Plants achieve a relatively high rate of assimilation of weak light, and a site cut off from early morning or late afternoon sun is not fully compensated by the strength of midday sun.

Slope and Cultivation

The most obvious of the direct effects of slope on agriculture is in the limitation of cultivation, especially in the use of machinery. The ideal slope for cultivation of most crops is between $\frac{1}{2}$ degree and 3 degrees. A very slight slope is advantageous for good drainage but on land with above 3 degrees of slope (1/20 gradient) some difficulties with machinery are encountered, and above 7 degrees the use of a combine harvester becomes difficult. Eleven degrees (gradient 1 in 5) is about the limit for regular combining or two-way ploughing. A 15-degree slope may, however, be ploughed if there is sufficient turning space above and below, but such a slope would normally be left in grass for long periods. Other difficulties occur on 15-degree slopes, such as in loading trailers, which may then only be removed downslope.

Some slopes of over 20 degrees in the British Isles are ploughed but only with considerable difficulty and some danger, and costs of liming and fertilising are high, so such fields are normally under permanent grass. At 25 degrees (gradient more than 1 in 2) normal cultivation methods are impracticable, grazing animals form paths across the slope and soil movement leads to marked erosion if good cover is not maintained.

Cultivation can be carried on without detriment to the soil cover on slopes in Britain that would, if cultivated in drier conditons, be likely to suffer from serious erosion. It is, however, technically possible to cultivate slopes steeper than are commonly attempted in Britain if discs rather than ploughs are used to break

[1] Garnett (1937), 52. See also Peattie (1936).

up the ground. In New Zealand slopes of 25–30 degrees have been cultivated with discs, though they are generally maintained subsequently in pasture. The most practicable way to improve such land is through fertilising and reseeding by aircraft, a practice already noted as used in New Zealand. In Britain, steep slopes are cleared of bracken by spraying from helicopters but subsequent costs of fertilising are too high — owing to the price of fertiliser (1976) rather than the aircraft operator's charges — to be economical for hill sheep and cattle grazing.

Steep slopes are generally restricted in pastoral use to sheep and cattle intended to produce meat, wool and hides because dairy cattle yield well only on flat or gently undulating land. Goats are kept in some areas but they cause severe erosion through their eating habits even on moderate slopes and their keeping should be discouraged. Sheep, also selective grazers, also cause erosion, while cattle contribute particularly through their treading but have less overall effect.

There is, unfortunately, no international or even national agreement on how slope should be classified for descriptive purposes but the Soil Survey of Great Britain has adopted slope classes, as shown in Table 2, based largely on the difficulties encountered in using machines for cultivation and farm transport.

TABLE 2

SOIL SURVEY CLASSIFICATION OF SLOPES

Degrees

0– 3	Gently sloping
3– 7	Moderately sloping
7–11	Strongly sloping
11–15	Moderately steeply sloping
15–25	Steeply sloping
25+	Very steeply sloping

Source: J. S. Bibby and D. Mackney, *Land use capability classification*, Soil Survey, 1969.

This classification would appear to be suitable for general adoption subject to the elimination of the overlap in classes by the terminology 'Under 3 degrees; 3 and under 7', etc.

Another problem arises out of confusion between the single slope, to which the above classifications are directed, and the compound slope. Table 3 illustrates both the difference in concepts of steepness, and the complications that arise when compound slopes are considered.

TABLE 3

CLASSES OF LAND FORMS ACCORDING TO COMPOUND SLOPES[1]

Symbol	Name	Slopes
0/	Flat	
1/	Flat to gently undulating	
2/	Easy rolling	Most slopes under 5 deg.
3/	Rolling	Most slopes under 12 deg.
4/	Moderately steep	Most slopes under 23 deg. many between 12 and 23 deg.
5/	Moderately steep to steep	Most slopes under 30 deg. many between 18 and 30 deg.
6/	Steep	Many slopes between 30 deg. and 38 deg.
7/	Very steep	Many slopes of 40 deg. and over

Run-off and erosion vary according to steepness of slope, and experiments have shown that, by and large, the erosion per unit area increases 2·5 times as the degree of slope is doubled.[2] During run-off, water accumulates as it flows down a slope. Consequently, more water flows over the lower part of the slope, and it flows faster than it does over the upper part of the slope. As a broad average, soil loss increases 1·5 times per unit area when the slope length is doubled.[3]

Although liability to erosion is always greater on steeper slopes than on gentle ones, this liability does not increase at the same rate for all soils. Thus, on a slight slope sand may erode less than clay, but on a steep slope the sand may erode more than the clay. Table 4 shows that ease of detachment of sand particles becomes more important on a steep slope than the ease of transport of clay particles.

TABLE 4

EFFECT OF SLOPE ON DETACHMENT AND TRANSPORTATION[4]

| Soil type | kg of water required to move 1 kg of soil | | Soil characteristics | | |
	Slope 8 per cent (4½ deg.)	16 per cent (9 deg.)	Infiltration capacity	Detachment hazard	Transportation hazard
Sandy loam	179	7	High	High	Low
Silty clay loam	65	24	Low	Low	High

To control erosion and facilitate cropping on slopes various forms of terracing are used. Found in many countries are bench

[1] New Zealand D.S.I.R. (1962), 30.
[2] Kohnke and Bertrand (1959), 103.
[3] Kohnke and Bertrand (1959), 104.
[4] Kohnke and Bertrand (1959), 110, quoting Duley and Hays (1932).

terraces, formed by building walls across a slope so that the soil can be accumulated upslope at a lower angle of rest than the natural slope. Stone-faced terraces in Peru date back probably four thousand years, while some in China, Japan and the Philippine Islands are over two thousand years old. Other old examples are found in Mediterranean countries, and more recent ones in the United States of America. In the U.S.A., however, they are no longer built because run-off water is difficult to control and erosion may continue to be serious.

Irrigation terraces are constructed by levelling sections of the slope and building retaining walls. Water is conducted from terrace to terrace by a system of flumes and weirs. Such are the well known terraces of south-east Asia, permitting wet rice and other crops to be grown on quite steep slopes, with almost no erosion. The Philippines provide the most spectacular examples. In northern Luzon terraces have been built to altitudes of over 1500 metres. In some cases they are only 3 metres wide, separated by almost vertical banks of up to 15 metres in height. Irrigation water is carried in bamboo conduits for two or three miles in some cases.

Terraces such as these are not suitable for mechanised agriculture. If large machines are to be used terraces must be broad-based with drainage channels. These are widely used in the U.S.A., but are difficult to build and utilise with heavy machinery on slopes of more than about 8 degrees. With mechanised farming, terracing is mostly on land of less than 5 degrees slope, and, where slopes are long and climatic and soil conditions facilitate rapid erosion, terracing is undertaken on slopes of less than 1 degree.

CHAPTER 3

Social and Economic Factors influencing Agriculture

It is obvious that the development of a particular type of farming in any region will depend not only on the land and climatic conditions but on the farmer's perception of his opportunities and limitations, economic and social, as well as of possibilities for the future. The many complex factors that need to be discussed under the label 'social and economic' or 'human' are less easy than the physical factors to group systematically and the operation of one is less easy to distinguish from that of another, partly because it is difficult or impossible to test the operation of any of them with the scientific methods which can be applied to natural phenomena.

In this approach to the problem, consideration will be given first to the social question of the nature and status of farming people, the difficulties facing the development of their societies and the influence of land tenure on the economic opportunities that are open to the farmer. Apart from the land resources at his disposal and social or religious taboos, the decisions of the farmer on what he shall produce are dependent most closely on the opportunities for marketing produce, for, except in the case of subsistence agriculture, there is no point in producing anything that cannot be sold. Effective marketing depends on transport, so this is considered next. Marketing is, however, hindered and transport improvements negatived, by economic controls, notably by tariffs and import restrictions, which are then discussed. These external factors will, however, be interpreted in different ways by different individuals, and their effect will indeed vary according to land and climatic resources within any one country to which they apply. So the choice as to the type of enterprise undertaken will be influenced by all the foregoing factors, together with personal desires and ambitions which are reflected in the level of income aimed at in practice, and which may be the deciding factor in the type of farm that is evolved in any particular situation. In evaluating the financial factor it is often overlooked that it is the income and cost situation at the margin that should, in economic terms, be decisive, and in the discussion of this economic concept it is appropriate to consider the concept of marginal land.

All the factors already mentioned are fundamental but the nature and amount of land at his disposal, and the external economic conditions under which he operates are only to a very limited extent under the control of the farmer. He can influence these in only the most minor and indirect of ways because they depend on society as a whole, international as well as national. Two inputs which are more under the control of the farmer are discussed next, labour and machinery. These are each variable according to the decisions of the farmer and they are to some extent interchangeable, the choice between them being usually made partly on social and partly on economic grounds. One could extend the consideration of economic influences to their impact on other inputs, such as seeds, fertilisers, irrigation and drainage works, buildings, livestock classes and quality, and all manner of items, but space necessitates limitation to the more fundamental aspects which are less liable to adjustment in short-term economic fluctuations.

The chapter closes with consideration of geographical results of the operation of all these factors, added to the physical constraints, in the form of specialisation of area and the evolution of distinctive types of farming.

Social Hindrances to Agricultural Efficiency

Farming people vary greatly in knowledge and education, status and wealth from one area to another and, of course, within any one community. They include entrepreneurs and business-men with considerable resources; managers selected for their professional ability; owner-occupiers of every degree of wealth and knowledge, skill and determination; share-croppers (métayers) whose limited finance compels them to carry on their farming in a shared business and hired workers who have neither the resources nor, perhaps, the wish, to become farmers on their own account. Perhaps still the most numerous, however, is the class called the peasantry. This also is a term embracing widely different people, from western European farmers who own small farms and differ little from their neighbours (who may not be thought of as peasants mainly because they own more land and bigger herds) to landless rural workers in South America and people in Africa still living in tribal communities.

At one time it was common for learned writers, including those who studied land and society in depth, to eulogise the peasant and his way of life, because of its supposed simplicity and closeness to the land and nature. More recently, however, the tendency has been to see the flaws in peasant societies, and to see the very nature of peasant ways of life as inimical to economic and even to social progress. Even the belief that a peasant community would enjoy the virtues of sharing resources and willing co-operation in labour on the land is no longer accepted uncritically. True, in many peasant communities, there will be an obligation on people to turn out for such major seasonal tasks as sowing and harvesting *en masse* or at least to join together for certain tasks, but research has revealed the tensions that exist in most communities run on traditional lines and few societies are without problems of getting the younger people, especially those with the better education that fits them for urban or other salaried employment, to stay on the land and fulfil the same functions as their forefathers. As subsistence economies have given way to increasing importance of trading relationships and contacts have been widened, so the pressure on peasant societies has increased.

Among the obstacles to improvement are poverty, malnutrition and isolation out of which are born apathy and minimal aspirations, vulnerability to natural hazards such as drought and flood, suspicion of the more progressive within the community as well as of advisers from outside and the new methods they proclaim, internal disunity, dependence of decisions on groups and families, hindrances from class and caste relationships and the lack of mobility they impose and religious attitudes.[1] The latter particularly affect livestock farming, most notably in the Hindu preservation of cows and the Muslim prohibition of the pig as unclean.

Many peasant societies have also suffered from the loss of rural industries, through competition from factory goods. This

[1] Grigg (1970) gives examples of each of these obstacles to development.

'agriculturalisation' has reduced incomes and opportunities and worsened the crisis in countries of low agricultural productivity such as Italy and India, and Franklin argues that rediversification is necessary.[1] It should not, however, be used as an excuse for lack of land reform or education in land use.

Development of Peasant Societies

The majority of peasant societies urgently need carefully planned development programmes. Franklin proposes a simple classification of peasant societies based on the types of development or reforms they need.[2] In the first group he sees the peasants of western Europe who, though disrupted by the process of industrialisation, have shared the benefits arising from it. They are now subordinate to richer industrial sectors which can provide subventions for their reorganisation and improvement. The collectivised peasants of eastern Europe and the U.S.S.R. are on the fringe of this position, after having themselves borne much of the cost of industrialisation.

The second group is of peasant societies which are in a much worse position owing to agriculturalisation, the growth of population and lack of a strong industrial sector which could provide them with subventions. The large groups of peasants in Asia and the Middle East and, provisionally, the mass of those in Latin America would come in this category.

The third group comprises those who are not sedentary cultivators, or who lack many of the elementary features of peasant life, but who may acquire the features of a modernised peasantry. Franklin sees the plans of western experts as providing the conditions for this change, but it may be inferred from the discussion of subsistence agriculture later in this book that this change is likely to take place more and more widely whether or not consciously planned. Land shortages, population growth and negligible industrial sectors are features of some of these societies.

This third category are folk culturally on the edge of the peasantry, having not yet experienced social and economic reform to bring them to the stages of the first two groups. At this

[1] Franklin (1962).   [2] Franklin (1962), 11.

extreme, many authorities would not refer to the non-sedentary folk as peasants at all, but rather as tribesmen.[1]

Three features have been suggested[2] as differentiating peasantry from tribes and these at the same time summarise important aspects of peasant life and outlook. In the peasant,

(1) Landownership and inheritance become very important. Property, savings and investment enter into all calculations.

(2) Domestic animals, where they are part of the system, make man a slave to time as well as to property.

(3) Individual motivation and competition begin. In tribal society individual competition is generally not prominent.

Evans points out that ownership of land is not an essential characteristic of a peasant society — indeed serfdom was the lot of generations of peasants from ancient Rome to pre-emancipation Russia — but preoccupation with land ownership or rights may be said to be a peasant characteristic and the idea of personal and inheritable property is usually well developed.

Many changes in peasant societies come from above, i.e. from estate owners or governments, usually the latter, because it is often in the interests of the landowners to preserve the *status quo*. In the case of the English enclosures this was not so since economic conditions favoured change being imposed by the landowners. In Soviet Russia and eastern Europe government intervention was made to implement the tenets of the Communist Party as well as with the intention of increasing agricultural efficiency. In Ireland, land reform was imposed by the British government when long-continued agitation from the exploited peasants could no longer be disregarded.

Ireland affords an example of land redistribution in the spatial arrangement of holdings as well as in the ownership of land. This was achieved without compulsory removal of people from the land — a feature associated with adjustment to sheep farming in the Scottish Highlands. The form of occupation of the land in Ireland was the same as that prevalent in Scotland, though in Ireland it persisted later. Known as 'runrig' in Scotland and 'rundale' in Ireland, it was basically an openfield system, but quite different from the manorial openfield systems of England and

[1] For example, Evans (1956).   [2] Tax (1956), 421.

Europe. An 'infield' comprised all the easily cultivated land and received almost all the manure. The 'outfield' was used mainly for grazing but patches were dug over for crops from time to time in some sort of rough rotation. Hill and bog land was grazed in the summer months. This was a system suited to a country where rough land and local changes in soil types abounded. The infield was divided into strips, and any one man would have several strips in different parts of the field, so that he shared in all qualities of land. He would hold shares on the mountain in proportion to his plots in the infield, these shares enabling him to pasture a certain number of sheep or their equivalent in other livestock. The system had many disadvantages, notably the fragmented nature of a holding. To make matters worse, the strips in some cases were changed annually to make doubly certain that no man then had an undue share of poor land, but in practice this meant that no one had much incentive to improve his own land. On the common or jointly-owned hill land all the livestock intermingled so that attention to improved breeding or ridding animals of disease was doomed to failure.

This system persisted in Scotland well into the eighteenth century, and in Ireland until at least a hundred years later. Investigation into land tenure in Ireland revealed the hopeless muddle that reigned.[1] The replacement of this system by one of consolidated farms was accompanied by colonisation of hill land, usually enclosed in long narrow stripes or 'ladders' running from the cultivable valley to the rough grazing. Some of the jointly-owned grazings were similarly divided. Many of these, however, remain to indicate something of the complexity that was associated with the rundale system. Thus, one mountain slope in Northern Ireland, comprising little more than 500 hectares, is shared by 23 lowland farms, the souming or grazing rights being allocated in fractions of 400. Two holdings have only 7/400ths each; the largest is 32/400ths. This latter share would provide grazing for about 60 sheep. Even in cultivated land some traces of the former confusion can still be found in fragmented holdings.

As the rundale system was eliminated, so it became possible for constructive action to be taken to transfer ownership of land from the landlords, many of them absentees in England, to the farmers themselves. Legislation to achieve this end was introduced in

[1] Devon Commission: Digest (1847).

1870, some twenty-five years after the Great Famine had stressed the urgency for reform. It was not, however, until after the Ashbourne Act of 1885 that much progress was made. This authorised the advance to tenants of the entire purchase money for their farms. From then until after the partition of Ireland in 1920 the process of breaking up the estates continued. Nearly all farms now are owned by their occupiers, subject to annuity payments. The small size of farm, shortage of capital, and lack of education of rural folk, both generally and in agricultural extension work, reduced the beneficial effects that might have been expected from such a comprehensive movement.

In many countries where peasants account for high proportions of rapidly increasing populations, reforms are urgently needed but weak or unstable governments fail to implement them. Indeed, the lot of the peasantry has worsened in recent years in many densely populated countries, such as India and several south-east Asian lands.[1] Formerly there were opportunities for accommodating the increase in population on unoccupied lands but there is little scope now for this remedy.

India illustrates the tragic manner in which the growth of population, poverty and hunger can thwart the technological successes of a development programme. A major breakthrough in food production was achieved in India following the adoption in 1966–67 of new high-yielding varieties of grain. The High-Yielding Varieties Programme, popularly known as the Green Revolution, facilitated diffusion of the new seeds and essential associated improvements in irrigation and distribution of fertilisers and pesticides. Within six years 28 per cent of the area under the five major foodgrains was under the new varieties. The better developed areas with the richer farmers benefited most and there were wide regional disparities, but average foodgrain output rose from about 80 million tonnes (1965–66 to 1967–68) to 101·3 million tonnes (1971–72 to 1973–74). Unfortunately, the growth in population and frequent failures of monsoon rains have eliminated most of the gains in *per-capita* consumption and, with slight declines in output, hoarding becomes common and the poorer, landless classes suffer first. Smuggling proliferates and enforcement of laws controlling food distribution is then virtually impossible.[2]

[1] Rosen (1975).　　　　　　　　　[2] Chàkravarti (1976).

Land Tenure[1]

High among the problems that face agricultural societies are those connected with land tenure and pressure for reform is rarely absent. Peasants who do not own land or whose farms are too small for prevailing economic conditions understandably see the roots of all their troubles in tenures that offer little or no security or are otherwise unsatisfactory.

Form of tenure will affect farm operations in many ways. Among the most important are:

(1) Length of time available for planning the development of the farm and profiting by the results.

(2) Extent to which investments in the holding could be realised if need arose.

(3) Whether the occupier is dependent solely on his own resources in exploiting the farm.

(4) How much income must be set aside to meet obligations in respect of rent, mortgage, etc.

(5) Preference for investment in the land as compared with investment in livestock or other movables, investment off the farm, or consumption — obviously linked with (2) above.

(6) Possibility of extending or contracting operations by purchase or sale of land or adjustment through letting.

There is a widespread assumption that owner-occupation is necessarily to be preferred to tenancy systems because of the incentive to improve a holding that is owned compared with one that is merely rented. It cannot be denied that there is a special and powerful incentive to preserve and develop an asset which is under the complete control of the occupier and may thus be devoted to the welfare of his family and heirs. Qualifications, however, need to be made.

If a man is to farm well he must also have sufficient free capital to invest in fixed and movable equipment for the farm, good seeds, fertilisers, livestock and other necessities for the full exploitation of the land. He must also have the knowledge required to exploit these assets, and be able to call upon technical

[1] The concept of land tenure must itself be defined. It is taken here to mean the system, or individual agreement, whether written or not, under which land is held or occupied. It is not restricted to the meaning which may be placed on the term in law. It includes all forms of tenancy and also ownership in any form.

and economic advisers to supplement and keep up to date his own knowledge. To develop his enterprise he will probably also need credit. A legal system or tradition that discourages subdivision may also be necessary for the best use of the land in the long term.

These are the advantages that a well-regulated tenancy system operating in favourable circumstances can offer. In many tenancy systems it is the custom for the landlord to provide buildings and some other items of 'fixed' capital while the tenant provides his own livestock, machinery and other movable capital. This means that a tenant who starts off with limited capital can concentrate on adequate supply and quality in these assets without the problem of financing either the outright purchase of the land and buildings, or the purchase of the one and the construction of the other. It may be easier financially to go on paying rent indefinitely than to raise a large capital sum. Obviously the balancing of these two alternatives depends on the level and control of rents, the availability of capital or credit, and the terms on which the latter can be obtained. In some societies it is virtually impossible to borrow substantial sums at a moderate rate of interest even with reasonable security. Farming is not a business distinguished for security based on earnings, nor is the person who seeks credit to set up in farming likely to have much else to offer as security, other than the farm assets themselves. Where occupation of the land is almost entirely by ownership, good potential farmers may be kept out of the industry entirely through lack of capital.

For a tenanted farm to offer a prospect attractive to maximum care in husbandry and improvement of land and fixed capital, however, there must be a sufficiently long lease. It is here that tenancy systems usually fail to meet the needs of good husbandry. A tenant who has a lease of 20 years or more has considerable incentive to make his own improvements to buildings, drainage and fencing, if the landlord is unwilling or financially unable to invest in these. Such leases are, however, rare. Shorter leases may have similar effects if there is full and reliable provision for a tenant on outgoing to derive adequate benefit for improvements carried out. This, however, can impose a burden on a landlord who may be called upon to make payments for his tenant's labour and investment which he would not have initiated himself, and might not be able to recoup from an incoming tenant. Raising a rent to meet compensation of this kind may be difficult, especially if government control is exercised over rents to prevent exploitation by landlords.

Very short leases and consequent insecurity for tenants are common. In Ireland the fear of landlords regaining control of farms has led to restrictions on long-term letting but these have allowed a pernicious system of short-term letting to flourish. This is 'conacre', whereby land is let for a period of less than one year. Very high rents are obtainable — in 1976 from about £70 per hectare for grazing to £200 or more per hectare for potato-growing land for the 11-month period — owing to the demand for land to supplement small farms. The system enables a farmer to adapt his holding to his immediate needs but there is strong temptation for a man who is working the land for so short a time to extract as much as he can from it and put back the minimum, and systematic rotation of crops is lost. Many farms are let piecemeal year after year by owners who do not want to work their land but will not part with it because they can get bigger profits through conacre letting.

A central problem in the economic development of many countries today is the rearrangement of farm boundaries to provide more efficient units, and in many cases to encourage the consolidation of holdings into larger and more economic farms. This may necessitate the reversal of previous trends. Inheritance laws and customs have led to subdivision of farms and over the centuries units have become extremely small in many countries. In some the opportunities of work in the towns have led to some degree of consolidation by those remaining on the land.

Recognition of the need for action to remedy both excessive subdivision and fragmentation of holdings has led to the adoption in many countries of schemes of agrarian reconstruction. The magnitude of the task in western Europe was indicated in a study prepared by the Food and Agriculture Organisation of the United Nations: 'It can safely be said that about 50 million hectares in Europe still wait for land consolidation, 5·7 million of which are located in the Federal Republic of Germany, 14 million in France, 10 million in Italy and a very considerable area in Norway'.[1]

---

[1] International Institute for Land Reclamation and Improvement (1959), 22–23. Clout (1972) gives details of a selection of European rural reconstruction schemes.

Given average conditions, farms of 10–20 hectares were considered viable in the Netherlands, Sweden, Denmark and West Germany in the 1950s but it could be argued that 100–150 hectares would be more appropriate in present-day conditions. The Mansholt Plan (1969) set targets for minimum sizes of 80–120 hectares for arable farms and 40–60 cows for dairy farms but these sizes have also been found inadequate. Progress towards these objectives has, however, been very limited and most western European farms remain hopelessly uneconomic.[1]

The value of many of the smallholdings is reduced by fragmentation. When fields are separated by land belonging to or farmed by other people, working efficiency is much affected, and there is increased scope for disputes. Modernisation, such as enlargement of fields and conversion to electric fencing or improvement of access to fields, is hindered or made impossible. Small size and irregular shape hinders cultivation by machinery and, where fences are absent or inadequate, grazing may be impracticable. Agricultural advisory work and soil conservation measures are impeded.

In other continents similar trends have been apparent and subdivision has been carried to extreme lengths in some countries. Muslim, Buddhist, Hindu and most African tribal laws provide for subdivision among heirs.[2] But, writing of paddy land, B. H. Farmer has argued that given the demographic circumstances and the lack of alternative employment, subdivision is better not discouraged.[3] Since paddy rice is a crop which responds well to intensive methods, and demands very careful water control which produces 'self-contained' units of land, 'there is no inefficiency in the sort of subdivision of paddy lands which goes on'. In any case, it is argued, legislation to control subdivision would attack only a symptom, not the disease, which is an economic system geared to a static population, but faced with a rapidly expanding one. Similar arguments are not applicable to tea and rubber lands, where economies of scale are relevant.

Communal tenure, still associated in the British Isles with some hill grazings and similar unimproved land, is common in less advanced societies. Many varieties exist, especially among the tribes of tropical areas. In some communities, although individu-

[1] Franklin (1971), 20 ff. [3] Farmer (1960).
[2] Webster and Wilson (1966), 95.

als do not own land, they have user rights which are permanent so long as they are exercised. In some cases there are individual user rights in arable land and communal rights in grazing land, which may extend to fallow land and stubble. Although it may be recognised that higher productivity should result from ending communal ownership this may be virtually impossible because of the complicated form of community rights, water distribution or difficulties of access.[1] The magnitude of the task and the risks involved in interfering with established traditions favour retention of the system but it must be blamed for widespread overgrazing and lack of improvement.[2]

Marketing

Apart from the purely subsistence economy, now of limited occurrence in the world and becoming rapidly more so, farmers will give priority in production to commodities for which there is an effective demand. This demand may be exercised by people in the locality, necessitating only the simplest forms of market, in which producer and buyers deal directly. At the other extreme, the consumers may be 20,000 km distant from the farmers, as is the British market for New Zealand produce, with a complicated machinery connecting the two groups.

Unless farmers develop some form of group control they are usually in a weak position in relation to their markets. In the simple market place buyers have the opportunity of dealing with many sellers and conditions approximate to perfect competition. If one seller is offering a standard commodity at a lower price than elsewhere, the others will not be able to sell their goods — unless there are non-economic considerations — until the cheaper stocks have been sold. This tends to lead to price-cutting and inadequate returns. Only by agreeing among themselves the prices at which they will offer their goods can the sellers hope constantly to obtain remunerative prices in conditions of balanced supply and demand.

Where an intermediary or merchant buys from farmers in order to resell in the markets, the farmer is again in a weak position. There are many more farmers than merchants and it is relatively easy for the merchants to dictate the terms on which they will buy. They know the market position better than do the

[1] Webster and Wilson (1966), 94.   [2] Grigg (1970), 119–122.

farmers, and can insist on their own assessment of a reasonable margin to cover their risks. The sum of the margins of each intermediary may'result in a very low price to the farmer (who has little alternative in selling) and a high price to the consumer or shopkeeper (who has few suppliers).

While individual merchants may adhere scrupulously to fair principles in their dealings and aim only at moderate profits, farmers in many countries have found it necessary to form co-operatives for disposing of their cash crops and livestock. The existence of a producers' co-operative inevitably has a salutary effect on the conduct of private dealers, who may still retain the allegiance of many farmers, but only so long as they maintain reasonable standards in their transactions. Some commodities lend themselves particularly to co-operative marketing schemes, dairy products being an example. Liquid milk must be collected from the individual farm and sold almost at once, while butter cannot long be stored except under deep-freeze conditions. Co-operative dairies solved the problem of the small producer in Britain and Ireland, not only by providing a ready market for milk, but also in being able to produce a standardised butter, which overcame the objections of variability in the product of the individual farm. Danish co-operatives pressed standardisation and grading of produce to remarkably high levels in the building-up of their national export trade.

Where controlling agencies are needed for marketing farm produce, but the producers have not themselves the necessary finance or skill to set up and administer co-operatives, government action is called for. Pre-war price falls and war-time problems led to such action in British West Africa and after the war Marketing Boards were formed to deal with the major export crops of this area. Government agencies bought from farmers and disposed of the crop at world market prices, retaining any surplus over the agreed price to build up funds from which local prices could be subsidised when world prices were low.[1]

The existence of co-operative agencies, producer-boards, and government bodies which carry out similar functions, has profound geographical significance. Without such organisation the production of a crop may be inhibited, even if the region concerned is well suited to it. Fluctuating market prices and

[1] See Pedler (1955) esp. ch. 17.

exploitation by commercial operators interested only in short-term gain mean insecurity for the producer. This may result in the decline of even a well-established farming system. If specialisation on cash crops produces disastrous results in several years the farmers will be forced back on subsistence or near-subsistence economies to the detriment of themselves, their would-be buyers and the whole economy. In contrast, a good marketing organisation can do much to ensure the economic stability of a region, the maintenance of a pattern of production suited to the geographical environment and steady development of the economy.

Transport

It is, of course, also essential to have transport systems able to convey the goods from producer to buyer. Development of new agricultural areas to keep pace with population growth in Asia, Africa and Latin America has been hindered by the difficulty of providing good communications. In spite of acute pressure on the arable land of the Andean republics of Bolivia, Peru, Ecuador and Columbia, potential farmland long remained untouched owing to lack of transport. Roads should be given high priority in all development schemes. The advantages of low unit cost and operating flexibility of road vehicles have been proved in developed countries. Railways, however, retain many advantages for sustained heavy traffic. Africa probably shows the closest relationship between railway provision and agricultural develop-ment, especially for export crops and staple foodstuffs such as cassava and yam, which are very bulky.[1] East Africa's bulk products, notably coffee, cotton, sisal and tea are very largely dependent on rail facilities.[2]

Many forms of produce require of their transport agencies more than mere capacity. Perishable produce demands speed and frequency of movement, or special measures for preserva-tion, or both. The dairying and meat exporting industries of Australia and New Zealand could never have entered world markets without the advent of refrigeration. Banana plantations, with their immense output, also must have high-capacity refriger-ated freight vessels for their economic operation. Special consign-ments of high-grade produce may be able to withstand the high

[1] Grigg (1970), 89–90.   [2] Hoyle (1973), 57.

costs of air transport if this enables them to be placed fresh on suitable markets. Selected lamb carcases have been flown from New Zealand to the London market by normal air-freight services, and consignments of strawberries have been flown from New Zealand orchards to Britain and Italy and peaches to Hong Kong.

Such developments suggest possibilities of long-range marketing by air, which have been improved by the introduction of aircraft capable of carrying more than 90 tonnes over 3000–4000 km stages. Large freight holds in passenger aircraft offer economical shipments of high value produce. Short-range air freighting is already well developed and includes, for example, the marketing of beef in northern Australia, and of fruit and flowers for Britain from the Channel Islands, Scilly Islands, France and other warmer regions. Again one sees the use of aircraft for perishable commodities. The Australian 'air beef' is exported frozen, the advantage of air freight being that the cattle can be saved the exhausting and time-consuming trek from inland stations to the ports, killing being carried out on the stations.

Air transport, however, is still used mainly in exceptional cases where speed of transit is essential or surface communications exceptionally poor. Even in local transport, however, speed is a factor where produce is perishable. Market gardening normally uses road transport in which lorries are owned by the farmers themselves or are closely geared by local transport agencies to their needs. As in other geographical patterns, road transport has made possible the diffusion of market gardening over a wider area than was formerly possible, though it is still noticeable that big cities retain near them areas of horticulture. In general the nearer the vegetable grower is to market the better are his chances of profitable operation, and the intensity of working makes possible the paying of rents which cannot be met by other types of farming (see Chapter 9). The producers of the slightly less perishable and more easily bulked commodity of milk have been able to supply the cities from greater distances through the media of special trains and tanker vehicles.

With less perishable goods, frequency of transport may be as important as speed. The less the time spent by consignments on the wharf, in loaded wagons or in transit sheds, the less the need for speed in actual movement, as well as the lower the costs. The farmer, or the market organisation supplying markets overland, has here a great advantage over those who must send by sea, with all the transhipping involved, as well as waiting for less frequent services. Thus the Irish producer of cattle or potatoes for the English market is at a disadvantage compared with his counter-part in England. The crossing of the Irish Sea means a longer time for calling forward supplies and greater risk of missing the best prices in markets with short supplies. Most services between Ireland and Britain are operated on six nights per week, but goods have to be alongside early enough for handling between rail or road vehicle and ship. The total time taken by a consignment of butter from an Irish factory to London is not much less than a Danish consignment takes. The disabilities in transport that have to be overcome in marketing New Zealand produce in the British Isles are referred to later (Chapter 5).

Transport must always be evaluated not only in terms of capacity, but also of cost. If transport charges are higher than production can bear there will be no incentive to produce for the market. Transport affects the farmer, of course, not only in the outward shipment of his produce, but also in the supply to the farm of seeds, fodder, fertilisers, store livestock and all other goods required for the farm and household. Transport charges almost always loom large in costs of agricultural production, and minimising these costs will extend the area of production for given markets.

Tariffs and Import Restrictions

While transport serves to extend the area within which produce may be sold, other forces operate to impose restrictions. These are the tariffs, quota restrictions and other import controls that are employed in one form or another by most countries.

Import controls are employed mainly to protect high-cost home produce from low-cost imports. Tariffs may be *ad valorem* or at varying rates according to the specific items of produce. Among the great trading countries which have recourse to tariffs to protect their own farmers are the United States of America and the countries of the European Economic Community. The E.E.C. countries have accepted the principle of eliminating tariffs between the member states and maintaining a common external tariff for produce from all other countries. The level of this tariff varies from low rates or zero for products which the E.E.C. countries cannot produce at all, such as tropical foodstuffs, to high rates for products of temperate regions which would compete directly with their own. Even with a high tariff to pay, low-cost producers such as New Zealand can market produce at prices which enable them to compete with the high-cost home product, so quota restrictions are used in addition. Quotas limit the amount of any one commodity which may be imported from any other country. They may be legally imposed, or accepted voluntarily by exporters who might otherwise face complete exclusion from the market.

At the time of expansion of world markets early in the nineteenth century Britain restricted the import of agricultural produce, but from the repeal of the restrictive Corn Laws in 1846 Britain followed a policy of free trade for nearly a century.[1] This enabled the country to draw supplies of food cheaply from the low-cost new lands of the Americas, Australia and New Zealand as well as from tropical areas. Home farmers suffered because they could not produce as cheaply on farms which were relatively small and inefficient, and therefore high-cost. With no other major international markets open freely to them, the overseas producers concentrated on exports to Britain and when world prices declined sharply after the World War of 1914–18, prices in Britain for agricultural produce fell drastically. The British government, therefore, was forced to abandon free trade and at the Ottawa conference of 1932 restrictions were accepted by exporting countries.

Restriction of imports, however, whether by tariffs or by quantitative measures, did not appear to successive British governments to meet the circumstances of an economy based on world trade and importing approximately half its food requirements. From 1938, when the increases in food production needed to enable the country to face a major war were sought, until Britain's entry into the E.E.C. in 1973, and to some extent thereafter, subsidies and grants were used to enable the home producer to obtain sufficient income to be able to sell his produce

[1] For a study of this period and subsequent agricultural policies in Western Europe see Tracy (1964).

in competition with low-cost imports. With few exceptions, such as temporary control of butter imports, when supplies offered from abroad were unduly high and prices consequently low, the subsidy system coped with the problem of enabling the British producer to command a good share of the market without shutting out foreign produce. Britain's entry into the E.E.C. and consequent adoption of its Common Agricultural Policy (C.A.P.), which tries to remunerate farmers adequately by maintaining high prices to the consumer, necessitated a gradual phasing out of this system, and the introduction of levies and tariffs on produce from outside the Common Market.[1]

At the same time there is considerable interest in many countries in reducing tariffs so as to stimulate world trade. The General Agreement on Tariffs and Trade (GATT) has been in operation since 1947 and there are complicated bilateral and multilateral agreements between countries both inside and outside GATT. Unfortunately, tariff reductions usually threaten the income of home farmers, and the strong force exerted by these farmers through their voting power makes reduction of tariffs hazardous for governments whose political future is not assured. Hence low-cost producers continue to be largely shut out from markets like the U.S.A. and Europe, even though this in turn severely restricts the amount the exporting countries can spend on purchases of manufactured goods and other needed imports. New Zealand is the classic case of a country which produces farm produce at extremely low prices in spite of a high standard of living, yet which, because this produce is denied more than very limited access to American and European markets, has to impose rigid limitation of imports. Uneconomic development of small scale industries follows in order to supply needs which could be met much better by the manufacturing countries, which in turn have to tolerate high food prices because of their own tariffs.

New Zealand is fortunate in being able to maintain a generally high standard of living in spite of these limitations. Such is not the case for India and other underdeveloped countries whose trade is limited by the tariffs and restrictions imposed by the wealthy manufacturing countries. Both industrial nations and many agri-

[1] For a detailed study of the effects of British entry into the E.E.C. *see* Davey, Josling and McFarquhar (eds) (1976).

cultural producers at all levels of development suffer from the national and trading-bloc policies pursued in the leading states.[1]

It is impossible to evaluate fully the effects of these financial controls on the geographical distribution of agriculture. It is, however, fairly obvious that if free trade in agricultural products were general, adjustment in the high-cost food producing countries would have to be widespread and would undoubtedly be painful to the agriculturalists. Most of the small farms of Europe might become amalgamated rapidly into much larger and more efficient units with a high degree of mechanisation to correspond more closely with conditions in the low-cost producing countries. As long as the small farms persist, sheltered by tariffs and supported by subsidies, a variety of crops and livestock may be maintained, some of which would disappear if truly competitive conditions prevailed. In the low-cost producing countries, access to wider markets at remunerative prices would encourage greater intensification of agriculture, though it is less easy to visualise the landscape changes which would follow there.

Political influences apply at the local as well as at the national level. The effects of state legislation and municipal land zoning ordinances on dairying in southern California provide an interesting example.[2] The northern milkshed, in the San Joaquin valley, had the lower production costs, but its advantage compared with the southern area, near Los Angeles, was neutralised by the prices set by the state agency. The state was divided into marketing areas and the prices established in each area were based on local production and marketing costs. The additional costs in the southern area were incorporated in higher prices. Restrictive zoning of land use to protect dairying was achieved by landholders taking advantage of state laws permitting any area with more than five hundred persons, with the consent of the majority of landholders, to incorporate into city status and establish its own zoning ordinances and local property taxes.

Type of Enterprise and Farm Income

Subject to special restrictions relating to plant or animal health and personal preferences, a farm will normally be devoted to the type of production which pays best — or which the farmer

[1] Johnson (1973).   [2] Fielding (1964).

anticipates will pay best — over a number of years. To the practical farmer, except in conditions of national emergency or to maintain fertility, there is no point in increasing costs unless income is likely to be increased by a greater amount. A decrease in income can be tolerated if costs can be decreased equally or more. The latter case was illustrated throughout Britain in the depression of the nineteen-thirties, when good arable farms were turned over to 'stick-and-dog farming' on land 'tumbled down to grass'. Sheep were carried on many farms at a density more applicable to a hill farm.

Small farms must be worked more intensively than large farms if their owners or tenants are to obtain reasonable incomes. There is a limit to which costs can be cut, and it should be worth a farmer's while to intensify his farming at least until he has eliminated under-employment of himself and any members of his family who are readily available to give him assistance. In Britain, for example, the most intensive forms of farming practised, apart from horticultural production, are pig and poultry keeping. Hence these have been favoured for the small farm.

The growth of highly specialised, large-scale units in both pig and poultry production, steep rises in the cost of feedstuffs and keen competition from suppliers have reduced the appeal of pig and poultry keeping to non-specialised farmers.

Dairying is the next most intensive enterprise, but about one hectare is needed for the year-round maintenance of a dairy cow, and it is generally judged economical and advantageous to maintain some younger stock for herd replacements on even a small dairy farm, so the size of dairy herd that can be maintained is severely limited. Some farms of as little as 10 hectares, with only four or five cows in milk, sell to the Milk Marketing Board, but it is only a generous pricing policy that enables such small units to remain in business, and they have greatly diminished in numbers in recent years.

Cattle rearing and fattening are less intensive enterprises, best suited to the relatively large farm, where cost of labour or indifferent soils militate against more intensive enterprises. Labour demands are low, except during calving periods, so area is usually the factor limiting the size of herd. At least one hectare of grass per head of cattle is needed for fattening purposes. Low rate of turnover makes it difficult to obtain an income on which to raise a family by this means with less than 40 hectares, able to fatten about 70–100 cattle per year in two drafts. Nevertheless, there are many much smaller farms subsisting mainly on cattle fattening at a low standard of living.

Cattle rearing, i.e. raising store cattle to $1\frac{1}{2}$ to 2 years old to be fattened on other farms, yields a lower gross income per hectare than the finishing process, even though more beasts may be maintained per hectare. Fattening needs feed of high nutrient level, which cannot easily be attained by farms on hill land, so concentration on rearing is usually more economic. Farms devoted to rearing probably ought to exceed 100 hectares in order to make satisfactory incomes, but there are many much smaller than this.

Arable cropping is also a form of farming of fairly low intensity and hence most suited to large farms. This was, of course, one reason why grain imported from the large farms of newly-opened America and Australia could easily undersell British grain in the nineteenth century. Potatoes, a labour-intensive crop, are more suited to the smaller farm, but the unbalanced demand for a great deal of labour for a short harvesting period discourages farms from devoting high proportions of their area to the crop, and the risk of disease from successive similar crops is another factor limiting their applicability on the small farm.

The growing difficulty which farmers have experienced in making satisfactory incomes from small areas of land has led many to seek supplementary incomes from other sources, such as catering for tourists. It has been said with some justification that in many areas caravans are the most profitable crop. Were it not for such possibilities and high subsidies many more farms would have to be amalgamated, but this process does continue steadily to reduce the number of separate farms.

Horticulture offers the most intensive use of land for food production and enables resistance to non-farming land uses to continue beyond the point that would be practicable with less intensive farming, but prices fluctuate sharply according to availability of supplies and the state of demand. This is a case where, given large enough holdings, the reverse argument may apply and less intensive production may be more attractive financially in the long run than the most intensive forms, and changes in this direction have been seen even in the south of England where land values have risen and demand for produce has grown rapidly during the past half-century. Other factors of production and marketing have changed, and with them there has been a tendency in Britain for intensive cultivation on small holdings with rapid successions of crops, heavy manuring and hand labour, to give way to a divergence of types of enterprise in vegetable cultivation. Highly intensive cultivation in glasshouses now contrasts with more extensive market gardening on a large scale using farm methods, as in the Fens and, more recently, in Bedfordshire. Conditions which brought about this change in Bedfordshire can be summarised[1] as:

(1) The end of supplies of cheap stable manure provided by London.
(2) The economic security obtained by growing large acreages of subsidised or guaranteed-price farm crops as well as vegetables for which prices fluctuate.
(3) The development of fragmentation on the holdings as some growers bought up plots from others.
(4) The introduction of Brussels sprouts, a hardy crop with a wide market as a winter vegetable, aided by motor transport.
(5) A labour shortage encouraging cultivation of easily-harvested crops such as Brussels sprouts, cabbage and lettuce.
(6) Economies of specialisation, including special low haulage rates for large consignments by both road and rail.

Different vegetable crops vary in the intensity of cultivation which is profitable partly because of their different labour requirements, which are indicated in Chapter 9.

Marginal Analysis and Marginal Production

For reasons such as those outlined above, the individual farmer has little control over the prices at which his produce will be sold. As we have already inferred, in order to obtain a surplus, or personal income, to cover domestic expenses, he will concentrate on production which will be marketable at prices which adequately cover his costs. If necessary, he must adjust his farming

[1] Beavington (1963).

to keep costs at an appropriate level. It is worth while to increase production by intensifying farming or increasing the amount of productive land only so long as the increase in revenue exceeds the increase in costs, i.e. marginal revenue exceeds marginal costs. The most profitable point in production is achieved when marginal costs and marginal revenue are equal, provided that total revenue is higher than total costs.[1] If production is increased beyond this point, using less favourable factors of production or involving loss of efficiency through the increase in size of the enterprise, gross income may continue to rise but net income (total revenue less costs) will fall.

Although the farmer may not think in these terms, as expressed by the economist, his notion of profit will be based on these inescapable facts. He may never attempt to calculate marginal costs or even average cost per unit of output[2] but if he is aiming to increase profits he will consider whether the employment of extra factors of production, e.g. labour, fertilisers, machinery, 'will pay' through yielding more than their cost. Although he has never heard of marginal analysis he may use the word 'marginal' to describe an expenditure which he thinks will barely 'pay for itself'.

The balancing of costs and anticipated revenue to achieve farm profitability is inevitably reflected in the landscape. The division of the land into farms of particular size and shape reflects past assumptions as to what comprises an economic unit; similarly with size and shape of fields, and the actual use of the land in terms of crops and livestock. This is not, of course, to say that the units will necessarily, in practice, be economic. They may not even have been economic when first laid out, because there has been a tendency throughout the ages and in all parts of the world for farms to be too small for satisfactory rewards to be earned. Furthermore, field arrangements which were efficient before the machine age may now be hopelessly outdated.

[1] For an explanation of marginal analysis and economic theory in general, see Samuelson (1973) or other modern texts in economics. Chisholm (1970) relates geography and economics, Morgan and Munton (1971) examine economic concepts in relation to agricultural geography.

[2] Average cost is lowest when marginal cost is still rising and may rise further before the most profitable point of output is reached, i.e. average costs do not directly indicate the most profitable level of production. However, the difference between average cost and average revenue per unit of production does indicate the level of profitability of operation and is more easily calculated than marginal cost.

Farming regions where successful adjustments have been made to changing cost-price relationships and changing income requirements are distinguished by an appearance of prosperity in the landscape. Where fields are well cared for, free of noxious weeds, soundly walled and fenced, with healthy crops and livestock and good buildings further characterising the farms, it is evident that the farm units are of economic size.[1] When costs are cut to preserve financial margins, maintenance of buildings is reduced, new outlays are avoided, and cheaper seeds mixtures, less labour, drainage work, etc. may be revealed in poorer land use.

For any particular crop or animal product certain places are particularly favourable for production, i.e. because of climatic, labour and other production factors, and convenient markets, income comfortably exceeds outlay. Away from these favourable areas, costs rise and returns are lower, until eventually the zone is reached where the particular type of production does not pay. Unless this is at the edge of a desert which has literally no use, some other product will be more profitable. Here then is the margin of profitability for two types of production — where it may be said that production of either commodity is 'marginal'. There may, or may not, be a clear physical margin coincident with the economic margin, but physical factors will enter into the position of the economic margin, even if only in the matter of physical distance from markets.

The margin of profitability for any particular product will shift areally according to the gap between prices and costs. Thus, given a rise in the price of sugar, and other prices not moving similarly, at the edges of a sugar-beet producing area where transport cost or less favourable soils have inhibited production, there will be an incentive for some land to be turned over to beet production. In many areas there is constant fluctuation of land use, reflecting a situation which faces farmers more or less permanently with marginal conditions.

It will be apparent that any land will be marginal in relation to some form of use, so care must be taken with the over-used term 'marginal land'. This term is often used when what is meant is

[1] Subject to the proviso that where the amount of labour on a farm is excessive in relation to production, i.e. over-employment exists, great care may be taken to preserve land and buildings, so giving an appearance of prosperity which belies the facts. Even in this case, the lack of new investment should be apparent.

simply land of low productivity, i.e. where yields are low in absolute terms or in relation to input. Such land is genuinely marginal if its use fluctuates, or if it is likely to be abandoned from all use because it is so poor. If it is in established use, as on an extensive basis for rearing range cattle or sheep, and there is no prospect of this use changing, it is not marginal land.

In the British Isles, for example, hill farming land is often incorrectly described as 'marginal land'. Some hill farming land is marginal in that afforestation presents an attractive alternative to its use for hill sheep and cattle, and at the lower edges of the hill farms there will be land on which the growing of crops, or the sowing out of improved pastures will be economically as well as physically marginal. Above the practical limits for afforestation, however, the hills are marginal only in so far as there may be competition between sheep farming and hunting interests. Possible fluctuations in use between deer, grouse and hill sheep, at perhaps one sheep to ten acres, does not seem to call for description of the land as marginal. It may be relevant that all these are extensive uses, whereas a more intensive use is represented by forestry or cultivation. It is where there is fluctuation or likely change as between such uses, and therefore between more intensive and less intensive categories of use, that the term 'marginal land' and its implications seem justified and desirable.

It is somewhat easier to be specific about what constitutes a marginal farm. Here the emphasis is firmly on the financial aspects of the farm unit. A farm should be regarded as marginal when it is unable to yield regularly a satisfactory profit, e.g. at least a worker's wage for the farmer plus interest on the capital, after making allowance for housing and other perquisites.[1]

Confusion, however, may arise because a farm which is inherently marginal by this definition may not include any marginal land, in the sense discussed above, while a farm comprising wholly marginal land may not be a marginal unit. The crux of the problem of the marginal farm is generally size. However poor the land, as long as it is capable of raising some crop or animal for which there is a demand, sufficiently large units will be profitable. Thus, there is nothing marginal about

[1] This definition was employed in a survey of marginal farms by the Department of Agriculture for Scotland (1947) and accepted by Ellison (1953).

Australian stations supporting one steer per square mile because the stations comprise thousands of square miles. The only marginal feature on a typical Scottish hill farm of 3000 hectares supporting 2000 sheep and cattle, is probably the lowest and best land on which it is barely economic to raise fodder crops. Such a use might not be economic if the farm could rely on purchased supplies of winter fodder. The whole farm would, however, become a marginal farm if prices for lambs and wool fell seriously, because the land would not permit of any other adequately remunerative use in farming.

Although, as noted above, marginal land and marginal farms are not always coincident, much marginal land is grouped in marginal farms, and many marginal farms are in their precarious financial state because their land is marginal and they have too little of it. This can be seen most readily in terrain which is difficult for farming and which has been colonised by people whose resources have been slender and whose efforts to make viable farm units have been frustrated by the lack of sufficient acres to offset the poor quality. There must be few countries which cannot provide examples, and all too often one may see adjacent areas of poor and better land with the farms on the poorer land smaller than those on the better. Many examples occur in Scotland[1] and Ireland.[2]

Labour

The labour available is a further important factor in farming. Labour requirements of different crops and different classes of livestock vary immensely. In typical British conditions, for example, about 5 man-days work are required annually per hectare of wheat, about 25 man-days per hectare of sugar beet and about 170 man-days per hectare of hops. Further examples are given in Table 8 (p. 216). Here we may merely note the great range in requirements, and consequent limitations on types of farming where labour is short. Scarcity of labour in newly-settled regions such as Australia and New Zealand and the western territories of America in the nineteenth century restricted agriculture to forms and methods which demanded little labour in relation to the land area available, such as wheat growing —

[1] Darling (1955), 202.   [2] McHugh (1963).

making maximum use of machinery — and stock ranching. As population built up, more intensive farming became possible and desirable, permitting the closer division of the land into smaller holdings.

At the other extreme, where there is a dense population living on the land, intensive forms of agriculture, capable of providing maximum subsistence, are necessary. The best-known crop of this kind is paddy or wet rice which can absorb over 4000 man-hours per hectare per year, or, say, the work of two people for 313 days per year for roughly $6\frac{1}{2}$ hours per day, and feed a dozen people. No crop can progressively absorb labour input without encountering diminishing returns, however; even paddy has an optimum labour input in terms of yield per worker well below that at which it is cultivated in many densely populated countries. As long as the labour cannot be better employed elsewhere, because of lack of alternative employment as in typical non-industrialised countries, and some additional food is being produced by the extra hands, there is justification for such disguised under-employment until the developing economy can take up the slack. If labour input continues to be increased, however, diminishing returns will eventually extend to the point where the marginal return of labour is zero. Clark and Haswell quote[1] some interesting examples of this situation. They note that the Japanese Ministry of Agriculture publishes rice yields and labour inputs for the 46 prefectures of Japan, and there is no discernible statistical relation between these two. Labour inputs range from 1400 to 2500 man-hours (including female labour) per hectare per year, all apparently above the economic limit. This result suggests the conclusion that above this limit additional labour inputs yield, in general, no return. Clark and Haswell further quote the analysis by Maruta[2] showing that output of oranges and tea on 30 small farms in the province with the lowest agricultural income per head in Japan can be almost entirely explained in terms of the inputs of land and capital. With average labour input as high as 2700 man-hours per hectare per year, the marginal product of labour appeared to be only 50 yen per day, or $0 \cdot 075$ kg wheat equivalent per man-hour, the farm price of partially-milled rice being 67 yen per kg. For 2700 man-hours this gives a total wage equivalent of $202 \cdot 5$ kg compared with an

[1] Clark and Haswell (1964), 89–90.   [2] Maruta (1956).

estimated general subsistence requirement of 250–300 kg of grain equivalents per person per year.[1]

Such very low marginal productivity of labour can be tolerated only where labour is not hired, unless wage rates are to be correspondingly low, and below subsistence level. Clark and Haswell find an appreciable amount of evidence in support of de Farcy's contention[2] that three kilogrammes of cereals is a normal reward for a day's work in a subsistence economy, examples being quoted from eighteenth-century France, nineteenth-century Belgium and Ireland, and surveys of conditions in Africa, Guatemala, India and elsewhere in the 1930s and more recently. In commercial agriculture, wages of farm labourers must be reasonably close to those available in other unskilled occupations if hired labour is to be obtained, and an appropriate marginal productivity of labour must be achieved if the employing farmer is also to receive due rewards. Difficulty in achieving this is an additional reason for specialisation in agricultural systems with low labour content in countries where labour is scarce.

Mechanisation

The effects of mechanisation of agricultural operations may be grouped as (1) displacement of labour and (2) extending the range of practicable operations.

In economic terms, replacement of labour by machinery occurs at the margin of profitability, i.e. when a farmer considers that investment in machinery, taking into account capital charges and depreciation, will increase profits through reduction in labour costs. Mechanisation may, of course, be stimulated by acute shortage of labour, as in the 'new' countries to which reference has just been made, but the economic rule holds good because the immediate result of shortage of labour is high wages for such men as are available. Mechanisation does not invariably reduce the demand for labour, since more intensive operations may become necessary, and absorb manpower on extra tasks, but the required skills change, with emphasis on mechanical knowledge replacing experience with simple implements and draught animals.

The history of agriculture has been a succession of examples of

---

[1] Clark and Haswell (1964), 51.

[2] H. de Farcy, *Revue de l'Action Populaire*, April 1962, quoted by Clark and Haswell (1964), 91.

SOCIAL AND ECONOMIC FACTORS 85

the extension of cultivation with improved tools, from the
introduction of the first light plough, referred to in Chapter 1,
and its successors in heavier and more efficient ploughs, to
horse-drawn implements and eventually, mechanically-powered
machines. It has been calculated that in England in the seventeenth century one acre could be ploughed in a day with oxen or
1½ acres with horses, but the steam plough of the nineteenth
century made it possible to plough 12 acres in one day.[1]

Perhaps the best way to illustrate progress during the last
century is to calculate the minimum number of man-hours
needed to produce a certain quantity of cereals from a given
area. In 1830 the production of 1800 litres of wheat on one
hectare, using the ordinary plough, harrow, sickles and flails,
took 144 man-hours. In the United States in 1896, with the
aid of the machines in use at that time, this had been reduced
to 22 man-hours; in 1930, when using tractors and combine-
harvesters, the time was brought down to 8¼ man-hours.
Thus between 1830 and 1896 a saving of 85·6 per cent in
time and 81·4 per cent in cost was effected.[2]

Many soils which were difficult to work, or on which cultivation
was practicable only during limited periods of the year so that
speed was essential, could only be brought under the plough
when suitable machinery became available. Examples may be
quoted from fertile but intractable areas within the world-famed
arable region of East Anglia to the arid regions of Kazakhstan.

The increase in output available for human consumption may
be noted from considering the economies of working land with
tractors compared with, say, a team of six or eight horses which
had to be cared for throughout the year and which would
consume the produce of a large proportion of the land they
cultivated. For this reason alone, cultivation of many lands of
moderate fertility became practicable only when the tractor
became cheap and reliable.

Areas which could not previously be cultivated for technical
reasons have also become more productive with improved and
specialised machinery. The example of steeper slopes becoming
cultivable as discs have supplemented or replaced the mould-

[1] Slicher van Bath (1963), 299.   [2] Slicher van Bath (1963), 300.

board plough was quoted in Chapter 1. Other examples are
numerous, such as the stump-jump plough which enabled land in
Australia not completely cleared of tree stumps to be brought
under cultivation,[1] and the mole-plough which makes possible
cheap drainage of clay and peaty soils.[2] Finally, brief mention
must be made also in this section of the use of aircraft for
top-dressing, sowing and application of pesticides, a topic which
will be developed in Chapter 5.

Specialisation of Area and Type of Farm

It has been indicated through references to income and
marginal analysis that it is the balance of costs and prices that
decides ultimately what commodities, out of the physically
practicable range, will be produced on a given farm at a given
time. Money can be viewed as a common denominator through
which comparison can be made of unlike commodities, and
similarly, if one can get the data, of the operation of unlike factors
in the physical and economic environment. This is not to say that
one can, or should wish to put a monetary value to every influence
in life, but, used intelligently, the financial yardstick is a valuable,
and, indeed an indispensable tool of the geographer as well as of
the economist.

Because what is economically advantageous for one farmer will
be attractive to his neighbours, with appropriate modifications
for size of farm and other variations in detailed circumstances,
there is usually a similarity in the operations of neighbouring
farms. A degree of uniformity is found throughout what we
may call a region (leaving consideration of this term and the
delimitation of such an area until later), so that we may speak,
for example, of a wheat-growing region, or, to refine this, a
spring-wheat region. A hundred miles away we may be in quite
different country which we may recognise as a cattle-ranching
region. Between the two there is likely to be a zone where the two
grade into one another — where the marginal revenue to be
derived from devoting an acre to cattle rearing is much the same
as that to be derived from spring wheat and its associates.

The localisation of industry, which became significant with the

---

[1] Wadham, Wilson and Wood (1957), plates 3 and 5.
[2] Watson and More (1962), 64.

industrial revolution, involved the separation of manufacturing, henceforth increasingly concentrated in factories, and agriculture, which itself became more specialised. This specialisation extended not only to concentration on farming to the neglect of domestic industry and crafts ('agriculturalisation'), but also specialisation within farming. By specialising, costs involved in a particular form of production could be spread out over a greater turnover, and economies of scale realised.

Specialisation of area makes easier the recognising of economic or agricultural regions, like the spring-wheat region and cattle-ranching region cited above, but regions may be distinctive without specialising. Thus, two regions may both answer to the description of mixed-farming regions, but production of a distinctive crop as a minor but recurrent feature of the farms in one of the regions may be the justification for distinguishing it from the other.

On the other hand, specialisation is not to be confused with monoculture. Areas of monoculture necessarily specialise but the reverse is not true. R. O. Buchanan makes this point clear[1] by reference to Denmark where 'we find a most beautifully designed pattern of an intensive long-rotation arable agriculture focused primarily on the support of dairy cows for milk, and through them on the production of bacon and eggs. There is no more highly specialised commercial agriculture area on earth, though it is by the system rather than by the product that it is characterised'.

The extreme specialisation of newly-won lands in the nineteenth century reflected the cost of the relevant factors of production. Land was normally plentiful and cheap, but labour very scarce and consequently costly. Extensive farming systems, with the emphasis on getting maximum return from labour, were the logical development. If sheep would survive and yield a marketable product with minimum attention, then sheep were likely to become the mainstay of the region. But as the population increased there would come both the economic incentive to subdivide large properties, and the labour to work the land more intensively and so yield a family's livelihood at probably improving standards on smaller areas of land.

Here we note that specialisation of area is a changing

[1] Buchanan (1959).

phenomenon. Not only do the internal circumstances of a region change, as in labour supply and rising land values, but also the external factors, including demand, transport, tariffs and the volume of produce from competing regions. It is all these things that are reflected in world commodity prices, and it is on these prices, as modified by local tariffs and subsidies, that the profitability of individual specialisation must be judged.

Even so, there is in any agricultural region a resistance to change, and it may well be that the more specialised it is, the greater will be the inertia — or the momentum — of the existing system. There is invested capital such as machinery, buildings and vehicles for specialised handling of particular crops or livestock, which is by no means limited to the farms, but includes the transport systems, commercial institutions and perhaps processing plants. Equally important is the accumulated knowledge of all the people involved, and the natural reluctance of most of them to make fundamental changes in their work before they have to. This is particularly so with the farmers themselves since (as often as it has been said, the truth of the statement bears repetition) farming is not simply an occupation but a way of life. For the sheep farmer, that way of life may well seem limited to sheep and a change to cattle or cropping harder to envisage than moving to a new country. This was exemplified in the mobility of the men who settled the new lands of North America and Australia.

Specialisation generally makes possible production of the selected commodities at prices lower than in more mixed production systems, because of the advantages of spreading the overheads of machinery and other capital inputs over larger quantities of output, as well as through the acquisition of knowledge and skills related to the particular enterprise. Grain farming, requiring large investments in machinery, and dairy farming, involving specialised buildings and equipment are examples, and where such specialisation is common a type of farming of a distinctive nature is readily recognisable, and if this becomes repeated throughout an area an agricultural region becomes identifiable.

Specialisation also makes for greater efficiency in marketing because the infrastructure can be built up to serve the needs of the producers, such as provision of grain elevators and dairy factories in the two cases cited above. Special vehicles for road and rail transport can also be employed with benefit to both producers and consumers.

There are, however, also some drawbacks to specialisation. Noteworthy among these is the greater risk of disease where there are large areas of the same crops or large numbers of particular animals vulnerable to particular pests and diseases. Good husbandry is also made more difficult by the seasonal concentration in labour demands and by the lack of balance which, in mixed crop and livestock husbandry, for example, stems from the interdependence of the two enterprises — the crops providing feed for the livestock and the livestock returning manure to the fields. There are also increased risks for the farmers who accept dependence on one or few products from the fluctuations of demand and supply. Because the demand for most farm products is fairly inelastic in the short term, large harvests usually result in lowered prices and, sometimes, the failure of the market to absorb the available supply even at sharply reduced prices. In this case, the specialised producer is likely to be badly hit and experience of such a situation often leads a farmer to some diversification as a form of insurance.

The lack of balance between demand and supply on the large scale is also often worsened by specialisation. Many farmers producing one product for the market may lead to chronic over-production. If left to free market mechanisms this would right itself but only at the cost of much financial loss to many of the producers. Furthermore, bankruptcy and a general with-drawal from producing that product might swiftly turn surplus into deficit. For political reasons it is generally not considered practicable to permit the free market to operate in this way and government intervention assures storage of the product in the hope that the market will eventually absorb it. The very fact of specialisation means that there is greater dependence on the one product than if each farm was more varied and the economic and political impact of financial distress makes it less likely that free market forces will be allowed to operate.

It is not, of course, only with specialisation that surpluses are produced. If financial returns for a product are expected to be good, and especially if security is offered to producers for the disposal of their crops through government agencies, if not sold in the market, it is likely that non-specialised farmers equally will seek to share in the profits.

The problem of surpluses is a serious and complex one. Fundamentally it means that farmers are being paid to produce unwanted goods. Not only is this a waste of money (usually for the taxpayers) but also a waste of resources — which ought to be devoted to a product more in demand, or conserved. It was not, however, this reasoning but simply the high cost of storage or immense losses in disposal, together with the damage being wreaked by over-cultivation, that led the United States government to adopt a policy of paying farmers to take land out of production.

Surpluses of crops may be used to supply needy areas in time of crop failure but there are negative aspects to such distributions. It has been argued that the ready availability of American food-aid has inhibited agricultural development in receiving areas, particularly through reducing the flow of products to cities from nearby farming areas, and so reducing investment in farming. Gale Johnson considers that 'Certainly food-aid contributes much less to the long-run improvement of peoples' welfare than making available an equivalent value in fertiliser, improved seeds and technical assistance'.[1] He noted that during the period 1954–70 the United States, under its PL480 aid programme, shipped abroad grain, dairy products, vegetable oils, cotton, tobacco and other farm products valued at $21,000 million.[2] Even if the humanitarian aspects of short-term assistance are considered to outweigh long-term effects on agricultural improvement and reconstruction in developing countries, there are economic and political problems for the donor countries in the cost of such a method of disposal of surpluses. Calls for a closer adjustment of supply with effective demand led to a change of policy to reliance on paying American farmers to withhold land from grain production. Meanwhile in Europe huge surpluses of dairy products and beef were being accumulated under the E.E.C. system of intervention buying and storage and disposal of these has ranged from selling butter in bulk to the U.S.S.R. at artificially low prices to the conversion of high grade foods into animal feeding stuffs — an extraordinary reversal of the normal production process. Apart from the sale of dried milk powder

[1] Johnson (1973), 164.   [2] Johnson (1973), 163.

there has been little benefit for needy countries in this situation but it has contributed substantially to inflation and other economic problems in Europe, which in turn have reduced the willingness of these countries to give aid to agricultural development in the Third World.

In these circumstances it may seem irrelevant to reiterate fears for the adequacy of world food supplies but it remains true that a large proportion of the world's population are undernourished and as the world population increases the problems are hardly likely to become easier to solve. Therefore planning by national and international agencies should seek to produce better balance of supply with demand, to avoid chronic surpluses and shortages and violent fluctuations in supplies with consequent gyrations of prices such as have characterised such basic crops as wheat, sugar, coffee, potatoes and rubber during the 1970s. Farmers undoubtedly like prices for their products to rise, but excessive increases depress demand and erode confidence. Stability, with prices rising gradually in conformity with long-term trends, suits them much better and permits the evolution of systems of agriculture which are themselves reasonably stable, evolving in response to technical innovations and the long-term adjustments of supply of land and labour rather than short-term, largely artificially induced, price changes. Even with such a desirable state of affairs, formerly approached, if not attained, by Commonwealth agreements of the kind that existed between Great Britain and New Zealand livestock producers and the West Indian and Australian sugar growers, systems of farming had still to be flexible enough to cope with the natural hazards of drought and storm and changes of taste and fashion. With these hazards governments should properly be concerned and short-term relief for farmers as well as for consumers is economically and socially desirable, but it should be made without perpetuating for decades structural imbalances in farm sizes, systems and output.

With all the forces for change that operate on the farmer it is perhaps surprising that the agricultural systems, types of farming and regions within which these are practised have remained as stable as they have during the past fifty years or so, but this is because the major divisions of agricultural typology relate to world climatic and soil zones upon which have been imposed organisational patterns which have become deeply entrenched, being altered only in selected details at any one time. Over time the total changes have been marked so that in temperate latitudes private enterprise farming has changed drastically from the old ideas of 'mixed farming' and collectivised agriculture is very different from what it was in the 1930s, while in the tropics great changes have been wrought in plantation agriculture and in many kinds of subsistence farming. Yet these organisational types of farming remain clearly distinguishable and established in the same regions of the earth as they were in the 'thirties. It is with both the durability of man's response to his total environment and with the detail of adaptation of his agricultural systems that the next section of this book is concerned.

PART II
SYSTEMS OF EXPLOITATION

CHAPTER 4

Agricultural Enterprises and Systems

The study of the effect of climate, soils, land tenure and other factors or forces on agriculture is analytical. The analysis is of complicated economic systems, the understanding of which is furthered by the analytical procedures. But these procedures are only tools for the dismantling of the mechanism. When it has been dismantled and its component parts examined it can be reassembled and its total form visualised more satisfactorily. Reassembly into a functioning whole, or synthesis, leads us to the recognition of the many different types of agriculture that exist in the world.

A farmer, influenced by these various factors, but making the final decisions himself according to his understanding of the forces with which he has to contend, adopts certain forms of work and production, or enterprises. Several enterprises are usually combined on a farm, the combination being called a type of farming. The word 'type' stresses the combination of similar attributes, but several interdependent types may form the whole complex system of agriculture of the wider region. Thus, store cattle and sheep grazing (one type of farming) occupies hill country, and stock fattening with cropping (another type) on lower ground, the two being integrated in an agricultural system, the former supplying store stock to the latter, which in turn may sell surplus feedstuffs to the hill farm for winter fodder but itself uses hill pastures in summer.

The areal extent of types or systems may or may not correspond with physical regions or political units. Hence, the identification and description not only of farming types and systems but also the areas, or regions, which they occupy, has long been a major concern of geographers. The methods of delimiting regions will be deferred until Chapter 9. For the moment we may concentrate on the identification of types and systems of agriculture and accept as a hypothesis to be tested later that these may be grouped spatially as regions.

Heavily committed as Great Britain was to the pursuit of international commerce, it is not surprising that, early in the existence of geography as a separate subject in British universities, its exponents gave more attention than had been given in other schools of geography to the classification and elaboration of the world's agricultural resources and products. The *Handbook of Commercial Geography* by George Chisholm of the University of Edinburgh appeared in 1889. Work by German geographers correlated the distribution of forms of agriculture and climatic regions, but a systematic approach to the classification of agricultural systems and regions based on the qualities of the agricultural enterprises themselves had to wait until much later.

From 1925 onwards a series of studies of agricultural regions, continent by continent, appeared in the American journal, *Economic Geography.* These added very substantially to the readily available knowledge of both world agricultural and pastoral systems and the regions within which they prevailed. In 1936 Derwent Whittlesey's classic study of major agricultural regions appeared.[1] The division of agricultural types by Whittlesey (with W. D. Jones) lacked attention to the principles of scientific classification but improved substantially on the grouping of 'forms of economy' which had been employed in earlier descriptions of world agricultural types and areas.

Whittlesey identified thirteen types of agricultural occupance,[2] with a further category for land totally unused for agriculture:

    (1) Nomadic herding
    (2) Livestock ranching
    (3) Shifting cultivation
    (4) Rudimentary sedentary tillage
    (5) Intensive subsistence tillage with rice dominant
    (6) Intensive subsistence tillage without paddy rice
    (7) Commercial plantation crop tillage

[1] Whittlesey (1936).
[2] Whittlesey used the terms 'type' and 'system' interchangeably, but it should be noted that a system is a functioning unit, which may be a single farm, or a group of interrelated farms. Farms of similar attributes which are not interrelated do not together form a system but are of the same type.

(8) Mediterranean agriculture
(9) Commercial grain farming
(10) Commercial livestock and crop farming
(11) Subsistence crop and stock farming
(12) Commercial dairy farming
(13) Specialised horticulture

Much criticism of this classification could be offered, and subsequent writers have modified the classes and the map which accompanied the original article (Figure 5). In a valuable review of attempts to construct schemes for agricultural regions Grigg[1] noted that although numerous articles on possible criteria had appeared, only Whittlesey, Thoman[2] and Kawachi[3] had actually devised classifications and mapped them on a world basis, and, further, that several of the completed and projected schemes appeared to have attempted to modify Whittlesey's scheme rather than to devise new systems. Grigg, while recognising the need for a reclassification of agricultural regions, stressed the problems inherent in the task. It therefore seems desirable to reproduce here the original classification, which has the advantage of stressing to the reader the origins of the work, which, though forty years old, still provides the basis of text book treatments of world types of agriculture and current atlas maps.

With this framework available, attention will then be given to several systems of agriculture which appear to the author to provide sufficient material to illustrate the approach to synthesis of the factors previously reviewed. It is necessary, however, to stress the interrelationship of the various factors and the dynamic character of any system. In a small book this can be done better by dealing rather more fully (though still only in outline) with a few systems with which the author is reasonably familiar than by attempting to cover in attenuated form even the main ones found throughout the world. These fuller descriptions (Chapters 5, 6 and 7) will not be dealt with under Whittlesey's headings. Indeed, it will be seen that each system overlaps several of Whittlesey's types, and the importance attached by the present author to the form of ownership and direction of the farms will become apparent. Whittlesey's classes, however, remain valuable as models against which any type of farming can be tested to reveal

[1] Grigg (1969).   [2] Thoman (1962).   [3] Kawachi (1959).

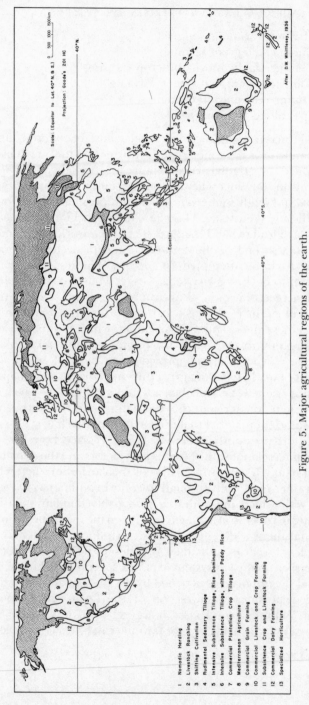

Figure 5. Major agricultural regions of the earth.

Source: D. Whittlesey, *Annals of the Association of American Geographers*, Vol. 26, 1936.
Compare recent detailed analysis of areas shown in Figures 6, 7, 9, 15 and 16.

1 Nomadic Herding
2 Livestock Ranching
3 Shifting Cultivation
4 Rudimental Sedentary Tillage
5 Intensive Subsistence Tillage, Rice Dominant
6 Intensive Subsistence Tillage, without Paddy Rice
7 Commercial Plantation Crop Tillage
8 Mediterranean Agriculture
9 Commercial Grain Farming
10 Commercial Livestock and Crop Farming
11 Subsistence Crop and Livestock Farming
12 Commercial Dairy Farming
13 Specialized Horticulture

Scale (Equator to Lat. 40° N. & S.)
Projection: Goode's 20I HC
0 500 1000 1500 km

After: D.W. Whittlesey, 1936

40° N.

40° N.

Equator

40° S.

40° S.

its characteristics as can be seen by comparing the following chapters with the descriptions given here with very brief additional comments.[1]

(1) *Nomadic herding* is the simplest form of pastoralism. Based on sheep, cattle, goats, camels or reindeer, it is primarily a subsistence form of exploitation of dry regions. The length of stay of the nomads in one place and the direction of movement are governed by the availability of water and natural forage. The habitation is a tent, a cave or snow hut, easily transported or replaced. Nomads have suffered reduction of their grazing areas as livestock ranching has moved into drier regions, and pressure has been put on many communities to adopt sedentary ways of life, notably in communist collectivisation of agriculture.

(2) *Livestock ranching* represents commercial use of physical regions similar to those of nomadic herding. Such regions in the Americas, Australia and New Zealand could be used only when herd animals and horses were introduced from the Old World. The movements of stock are normally confined to the ranch although there may be seasonal movement or *transhumance* to distant pastures. The ranch is a permanent base. Ranching methods are the outcome of European adaptation to environments demanding very extensive forms of agriculture. Cropping is limited to provision of fodder for periods when natural grazing is unobtainable or inadequate. The provision of better supplies of fodder, together with water control and improved care of stock are the principal ways of increasing the yield of these farms.

(3) *Shifting cultivation* is a primitive form of utilisation of the poor soils of tropical rain forest and bush areas. Farmed plots are moved every few years in search of accrued fertility, and dwellings are moved as necessary. Fire and hand tools are used for clearing the ground and for simple cultivation. This and the following types of subsistence cultivation are further described in a later section.

(4) *Rudimentary sedentary tillage* employs methods often not much more advanced than are found in shifting cultivation, and may arise from the settling down in favourable places of nomadic or shifting communities. Lack of fertiliser may necessitate fallowing of plots, and so the type is not sharply separated from

[1] For further comment and updating of Whittlesey's regionalisation see Grigg (1969, 1974) and Gregor (1970).

shifting cultivation, but, as the community develops its settled life, contact with commerce is likely to lead to cultivation of crops, especially tree crops such as cacao, oil palm and rubber, with which can be bought a few manufactured goods and food for the lean period before the harvest.

(5) *Intensive subsistence tillage with rice dominant* is, along with its companion type which excludes wet rice, the basis of life in south and east Asia. Deltas, floodplains, coastal plains and terraces to which water can be directed are utilised mainly for rice, which with its high yields per acre offers the best available subsistence to the dense populations. Land which cannot be so used is planted with other crops including tree crops. Fish supplement the agricultural produce. Work in the fields is arduous and unending and little assistance is available from machinery for tillage or irrigation, though ploughs may be hauled by water buffaloes and new small machines have been developed for use on smallholdings, notably by Japanese manufacturers. Whittlesey's emphasis on the subsistence aspect of the type does not reflect the growing influence of trade on intensive rice cultivation communities to which reference is made in Chapter 8.

(6) *Intensive subsistence tillage without paddy rice.*[1] Many regions of dense population which lack the conditions suitable for rice have types of agriculture differing from the foregoing mainly in the absence of wet rice. Some land is irrigated but more emphasis is placed on dry grains, and tree crops are also commonly important. South and east Asia and oases of inner Asia and north Africa, especially Egypt, are cited as the main areas of this type. Similar, but less intensive, methods are practised in the Kano district of the Western Sudan and the Mexican highland. Some modifications to Whittlesey's classification of subsistence agricultural types are suggested in Chapter 8, pages 170–172.

(7) *Commercial plantation crop tillage.* This alien system, introduced into regions suitable for production of tropical commodities needed by the industrial countries, occupies in the aggregate a very small area compared with the other types outlined. It is, however, important commercially and has resulted in the introduction into regions of subsistence agriculture of staff, methods, crops, machinery, fertilisers and finance from abroad.

[1] Whittlesey uses the term 'paddy' to mean only wet rice but this term is also employed for upland rice, or 'hill paddy'.

The principal crops are sugar, tea, coffee, cacao, rubber, oil palms, sisal and bananas. Most of these crops are also grown by smallholders. An example of this form of production is given in Chapter 6.[1]

(8) *Mediterranean agriculture.* Whittlesey argued that climate and mountainous relief led to a distinctive stock and crop association evolving in the Mediterranean basin; all-year or winter crops grown with rain, all-year or summer crops grown with irrigation, and livestock — mainly small animals — grazed on lowlands in winter and on highlands in summer. Tradition, market and government policies influence attention to particular crops, e.g. vines, citrus fruits, wheat. In the regions of Mediterranean-type climate in the other continents there are many similarities in type of farming, though in California wealth has made for many differences in irrigation and mechanisation, visible also in the corresponding regions of Australia. Later writers have much criticised the obviously climatic basis of this type and some have reclassified the areas under other descriptions.[2]

(9) *Commercial grain farming* arose with the availability of the steel plough, harvesting machinery and transport systems capable of cheaply handling export crops in bulk. Sub-humid continental climates with short summers, which could otherwise be commercially exploited only by livestock ranching, produce grain, mainly wheat, by extensive methods. Crop failures are common. The best known areas are in North America and Argentina, but mention is made in Chapter 7 of the Soviet lands of this type, which have been transformed by collective farms and state land development schemes.

(10) *Commercial livestock and crop farming* is also commonly known as 'mixed farming'. It is found in Europe, where it originated, and in the humid middle latitudes of all the other continents except Asia. Its development, however, is governed closely by marketing possibilities, with tariffs, subsidies, etc. influencing choice of crop or livestock. In the light of these and of climatic influences, varying emphasis is given to the different grains, root crops, cattle, sheep and pigs. Mechanisation, crop

[1] Gregor (1965) demonstrated the common features of the plantation and the large modern farm, extending plantation methods into temperate areas.
[2] See, for example, discussion by Grigg (1969), 124.

rotation and use of fertilisers are normal. The United Kingdom and New Zealand are chosen to illustrate variations in this type of farming in Chapter 5.

(11) *Subsistence crop and stock farming* resembled the foregoing type in some of the crops and animals found on the farms, but little or nothing was sold off the farm. At the time when Whittlesey wrote, this type was in rapid decline with collectivisation in the U.S.S.R. and, to a lesser extent, in other countries. These changes have now progressed so far that it is unimportant, and might disappear from a modern worldwide classification. A subsistence element is present, of course, on many farms, even large collective farms, as described in Chapter 7, but it is incidental to the main objectives.

(12) *Commercial dairy farming* also evolved in Europe but has spread to other regions which can take advantage of large demand from urban populations. It remains true, as Whittlesey wrote, that the radius for shipping fresh milk is, roughly, overnight; for cream twice as far; while refrigerated butter and many kinds of cheese can be supplied across the world. For marketing of milk in the stringent conditions imposed in the modern cities and for economic participation in the sale of graded dairy produce, high grade farming is required. Some further discussion of this type of farming appears in the sections on agriculture in the United Kingdom and New Zealand.

(13) *Specialised horticulture* has also developed in response to the large demand in urban centres for foodstuffs, but the most ancient districts of specialised horticulture are the vineyards of Europe outside the Mediterranean climate, some of which were established in Roman times. Burgundy, Champagne, the Moselle, Rhine and Loire valleys, the Swiss lakes, the plain of northern Hungary and the quasi-Mediterranean area near Bordeaux are the most important.

Urban demand for vegetables and soft fruit is satisfied partly from market gardens within a few hours' transport of the cities, partly by more distant 'truck farming' holdings. Market gardens near cities, where high land values obtain, are worked with special intensity, with high inputs of labour, fertiliser and other factors of production. Cultivation under glass increases intensity of work coupled with higher investment.

The more outlying truck farming regions exploit particularly favourable soil and climate, which enable production to be earlier in the season or at lower cost than in the suburban market gardens. Regions of this type of production are the Channel coastlands of France and the Low Countries, the Rhone valley, the north African coastlands, the south-east coastal plains of the United States and regions of drier climates further west, notably in Colorado and California. In some of these districts commercial fruit farming is important, with irrigation commonly employed, as in much of the vegetable growing. Specialised orchards occur also in areas of favourable climate in other regions.

In some districts of this type there is also attention to specialised cash crops such as sugar beet, tobacco and cotton. The irrigated cash cropping of some of the dry regions is similar to the vegetable and fruit growing of nearby oases. Specialised poultry production also resembles market gardening and truck farming in its organisation near large cities and in California.

In conclusion, Whittlesey comments on the cash basis of the farming types of the occidental middle latitudes and their dependence on the urbanised world created by the industrial revolution. This relationship is elaborated in the first of the following more detailed studies.

It should, however, be emphasised that the systems described in the following chapters should not be assumed to apply to areas of broadly similar type of farming or organisation in other parts of the world. Farming of the 'mixed' type in other regions may be very different from that of the United Kingdom and New Zealand — it will be seen that many variations occur between these two areas and within them, relatively small though they are. Plantations elsewhere reflect different conditions to those of Malaysia and state or collective farms in other countries may vary greatly from the Soviet system. These are merely examples to permit the development of principles and to illustrate recent trends in a world of rapid change.

CHAPTER 5

Mixed Agriculture — the United Kingdom and New Zealand

The term 'mixed agriculture' is used for brevity in referring to the combination of cropping with livestock rearing and dairying — covered by Whittlesey's types 10 (commercial livestock and crop farming) and 12 (commercial dairy farming). Mixed agriculture is usually understood to imply also a mixture of arable and grassland field cover on the farm though this is not always the case. Some livestock farms have little or no grass, whereas some all-grass farms carry several classes of livestock and may be regarded as mixed farms. The variety of combinations found in England and Wales is illustrated in Figures 16 and 17 and some of the variations in the more specialised grassland types in New Zealand are given in Figure 7.

Three main features characterise the mixed farming types discussed in this chapter:

(1) Several enterprises are found on most farms — often more than one type of crop and more than one class of livestock, with varying proportions of cereals, roots and grass contributing to both livestock feed and cash crops. Increasingly, however, specialisation is resulting in fewer enterprises and the traditional term 'mixed farming' becomes somewhat less appropriate.

(2) The objectives of the farming are commercial, i.e. the bulk of the produce is sold off the farm.

(3) The normal form of organisation is the family farm, and whether occupation is by ownership or tenancy the individual farmer is free to choose his system of farming, and he bears the ultimate responsibility for failure as well as for success.

The freedom of the individual farmer to make the strategic and tactical decisions is an important one. The fact that a very large number of farmers operating in a commercial economy should organise their farming to conform to a basically similar pattern without any form of direction is highly significant. Hence the division here of mixed farming of temperate latitudes into agriculture of this type, considered in this chapter, and agriculture of a somewhat similar kind, producing similar products but with state-directed organisation. The latter is discussed under the heading of state and collective farming, and affinities are seen between such forms and plantation agriculture. Plantations are organised by interests which are predominantly non-rural and typically not native to the regions concerned, usually limited liability companies which exist for the benefit of their directors and shareholders. The organisation of collective farms is intermediate in its representation of the interests of the workers on the land itself and the degree of non-rural control, though, of course, there are gradations between the different forms.

In reaching the high state of development noted above, the mixed farms of temperate regions have enjoyed two main advantages, both of great importance during the last two or three centuries. Firstly, it has been possible to grow a wide range of products with, in general, a fair degree of reliability. Improving technology has steadily widened this range of products and their yields in given conditions of climate and soil. Secondly, it is these products which have been most consistently in large scale demand. This demand has come from countries where rapid growth in wealth has accompanied industrialisation and expansion in population. Industrialisation results in an increasing proportion of people divorced from food production. Thus, there have been large markets in Europe for temperate foodstuffs produced in the 'new' lands of the Americas, the British Commonwealth and Soviet Asia as well as those grown in the European countries themselves.

THE UNITED KINGDOM

So great was the growth of the urban population of Great Britain during the eighteenth century that, in spite of new techniques and reorganisation of farms, production in Britain could not be expanded rapidly enough to provide the necessary food at prices that could be afforded by the factory workers. Hence the reversion in 1846 to free trade — the opening of the great British market to cheap foodstuffs from abroad. The full impact of this measure was not felt until railways, steamships and refrigerated vessels had revolutionised the transport of grain and livestock produce, but by the late 1870s British farmers were having to adapt to low prices or leave their farms. The succession of years of appalling weather from 1875 to 1882 worsened the crisis, which deepened into the depression which lasted until the end of the century. Ultimately, new methods of farming, in which costs were adjusted to prices and opportunities more efficiently exploited, emerged and set the pattern for the twentieth century.[1]

In the British Isles and western Europe, great extremes of heat and cold are rare, but extremes of water supply do occur. Husbandry of the land must have regard to the following outstanding conditions:

(1) The climate is particularly suited to the growth of grass. Grass will grow nearly all the year round, and, with appropriate harvesting of some of the grass for fodder, livestock farming may be carried out wholly on grass, and without irrigation. This is the feature that especially distinguishes these regions from others. Two major limitations, however, occur:

   (a) On the oceanic margins the high ratio of precipitation to evaporation (P/E) threatens grass through the development of weeds, and, ultimately, forest or bog.

   (b) On the continental margins, lower precipitation and high evaporation, associated with clear skies and long periods of sunshine, result in water shortage and scorching of the grass. As these conditions become predominant, cereals become more suitable crops than grass, but conditions also favour soil erosion so both agricultural and pastoral activity must be regulated accordingly.

(2) Climatic conditions on the oceanic as well as on the continental margins permit certain alternative forms of agricultural land use, based on arable cultivation. Cereals,

[1] Orwin and Whetham (1964) provide a comprehensive review, as they do for the whole of the period 1846–1914.

short ley pasture grasses, clovers, lucernes and roots such as potatoes and turnips are among the crops employed.

The response to these conditions is seen in the patterns of farming that emerge. At the one extreme is pastoral farming based wholly on grassland. Livestock are used to convert the grass into foods favoured by human beings, and into by-products such as wool, hides and horn. Sometimes these latter are the main products with foodstuffs subsidiary. At the other extreme is arable farming, producing cereal or root crops for direct human consumption or industrial use. More common, however, is some combination of these two forms, such as production of grass, cereals and roots to feed to livestock, any surplus being sold as cash crops. The result is commonly a mixed land cover — grass on some parts of the farm, and arable crops on other parts.

To which of the extremes any particular region will show a tendency, or what manner of compromise may be distinguished, will reflect the varying force of a number of factors which have already been discussed in general terms (Part I), namely:

(i) The local pattern of climate and weather.
(ii) The pressure of other physical factors — landforms, soil, etc.
(iii) The stage of technological development.
(iv) Historical and social factors which affect the response to environment and knowledge.
(v) The economic conditions.

Economic conditions are listed last, because they are at once the outcome of all the other conditions, and at the same time they sum up all the others. In other words, the objective of production is the product which in all the circumstances provides the greatest return for the input.

There are, of course, certain exceptions, where some other conditions override economic decisions. Thus, particular animals may be kept — or avoided — for reasons of religion, personal taste, or in accordance with regional traditions, but these are of very limited effect in the British Isles. Where commercial sentiments are strong, the economics of practical farming dictate that the farming shall be well within the climatic limits. Within these limits, the variations are in response to economic conditions which reflect the above forces acting in detail. As we examine the refinements in variations of agriculture within the broad climatic possibilities, so we become more and more aware of the operation of other factors. To sum up this argument, climate lays down the broad possibilities of the region; other forces decide how these possibilities will be exploited.

Europe shows admirably the transition from pastoral and semi-pastoral husbandry to concentration on grain between the oceanic west and the dry interior lands and, to a lesser extent, between the cold northern margins and the moderate mid-European latitudes. The transition can clearly be seen in its initial stages in the varying emphasis between pastoralism and arable farming in the British Isles. Because they have been occupied continuously during the evolution of modern mixed farming techniques, and were also the scene of many of the innovations that created the system, the British Isles offer particularly rewarding subjects for study. The adaptation of the environment to suit more intensive methods of production may likewise be studied.

Technological Development

British pastures, even of unimproved strains, will support large numbers of livestock in the summer but not in the winter when temperatures fall below the threshold temperatures for growth of about 5 deg. C. (41 deg. F.). Pastoralism alone, therefore, will not provide a healthy agricultural system. Harvesting hay provides a crude means of keeping stock alive but this is not sufficient for progressive agriculture. The growing of grain is marginal in the cooler, wetter west and north; yields of wheat are poor, oats and barley fare better, but neither from them nor from a combination of grains and hay could winter feed be obtained for all the stock pastured in the summer. Grain was grown in mediaeval England mainly for bread and beer, and there was little food value in the straw that could be spared for the animals. Hence the autumnal slaughtering of medieval times, which must surely have been considerable, even though not as drastic as has sometimes been supposed.[1]

As noted in Chapter 1 the introduction of root crops

[1] Trow-Smith (1957).

revolutionised British agriculture, but only slowly. Throughout the eighteenth century there was gradual improvement. Potatoes proved a bulky but fairly efficient food for both people and livestock. They tempted the Irish into too much reliance on them, so that famine followed the failure of the potato crops. Turnips were the basis of the agrarian revolution in East Anglia and spread throughout the British Isles as a basis for wintering stock, replacing fallow. The Norfolk system, a four course rotation,[1] was less suitable for the cooler and wetter regions, but was readily modified by the extension of the period in grass and replacement of wheat by oats or barley. In the north of Ireland, Atlantic outpost of arable farming, a sound rotation was found in oats, potatoes, oats and several years in grass.

Meanwhile grassland husbandry was put on to a scientific basis with the sowing out of pastures with the more nutritious grasses and clovers, the nitrogen-fixing qualities of which were fundamental to the maintenance of improvement. Hand-in-hand with the improvements in the feed basis went the development of the livestock breeds. Breeds of cattle, sheep and horses became increasingly differentiated and specialised, adapted to the differing needs of different regions and different types of farming. Thus, while general purpose breeds, such as the Shorthorn cow, were found to be extremely valuable throughout the British Isles, and have remained so to this day, this has not prevented their being developed into more specialised types like the Dairy Shorthorn and Beef (Scotch) Shorthorn. Regional breeds developed, particularly adapted to their local environments, such as the Ayrshire cow, capable of giving good quality milk in rather discouraging conditions of climate and relief, and the Aberdeen-Angus beef cattle. Interchange of breeds has facilitated adaptation to environment and market conditions. In cows, the Jersey and Friesian are good examples. The former supplies milk of high butterfat content, suitable for high class dairying in areas of high purchasing power, the latter gives good milk in great quantities and is a most valuable animal on which to base normal town supply.

---

[1] A root crop, spring-sown grain or peas, a grass crop grazed or cut for hay followed by autumn grain, usually wheat. Sheep were folded on the arable land and consumed the roots, providing dung, and cattle were fattened over the winter for the period of high prices.

Social and Economic Conditions

The new farming systems were possible only on land that was enclosed, and the pressure of growing population and food demands stimulated enclosure and redivision of the land. Land use is never static but there are periods when change is stimulated by sharply changing social or market conditions and technological development. In the eighteenth century, the influences demanding change were mainly internal — the growth of population and industrialisation — but this was nonetheless the impact of a widening as well as a more specialised economy. In the nineteenth century, the continuation of the growth in the economy was not paralleled by adequate growth in the ability of British agriculture to supply the quantities of cheap food needed in the cities. The supplies offered by newly developed lands abroad led to the freeing of imports. Competition from cheap overseas grain after the repeal of the Corn Laws in 1846 and subsequent transport improvements forced a reorientation of British farming towards livestock products which were protected by distance and the perishability of the products.

During the crisis of low prices in the 1880s another major adjustment was compelled by the introduction of refrigeration, making possible the shipping of meat and dairy products from the southern hemisphere. Fortunately for the British farmer there was a continually expanding market for fresh milk and vegetables. There was also a continued high level of demand for home produced meat, with its distinctive qualities. On these products farming could continue profitably as long as there was a generally high and steady level of employment, and consequently of effective demand.

The economic catastrophe of the 1930s showed the vulnerability of farming, like other sectors of the economy. Agricultural prices in all parts of the country plunged to half the levels obtaining in 1930, and for some commodities to as low as a quarter of the 1921 prices. The depression had not reached its lowest depths in 1932–33 when the land use pattern was being mapped by the Land Utilisation Survey of Great Britain[1] but the survey stressed the declining percentage of land under the plough. An increase in the proportion of grassland might not, in

[1] Stamp (1948, 1962).

itself, have indicated declining fertility, but it did in the circumstances of the time, the land being commonly allowed to 'tumble down' to grass, with consequent low standard of resulting pastures. Former arable farms were turned over to low-intensity grazing and buildings, fences and equipment deteriorated from lack of maintenance.

In the face of this crisis, the United Kingdom government resorted to a measure of protection. In 1932 an *ad valorem* tariff of 10 per cent was applied to some commodities. Under the Ottawa Agreements, exporting countries agreed to limit shipments of beef, mutton and lamb, and by 1934 quotas had been accepted by several foreign countries in respect of chilled beef, frozen mutton and lamb, processed milk and cream, potatoes, fat cattle, oats and eggs.

Whilst imports were being restricted, steps were taken to improve the marketing of home produce and so make for greater efficiency in home agriculture. The Agricultural Marketing Acts of 1931 and 1933 facilitated the development of orderly marketing of a number of products, with farmers themselves exercising a high degree of control. The Milk Marketing Boards were of particular importance because of the large and growing role of sales of liquid milk in the economics of British farms. Prices for milk were equalised, whether the eventual use was as fresh milk or manufactured products, so that farmers whose milk was used for the commodities also available cheaply from New Zealand and elsewhere were not penalised. This greatly helped farmers not located near cities and simplified the operation of the boards. Hops, bacon pigs and potatoes were the other important products handled by marketing boards but the latter two experienced problems in controlling the market and only the milk marketing boards were a major cause of changes in the geography of British agriculture.[1] The government also departed from its policies of merely providing assistance in matters of research, limited credit schemes and help in marketing by reintroducing subsidies, which had been paid during the First World War and then withdrawn. The first of the new subsidies was for wheat. Legislation of 1932 guaranteed a standard price and a secure market for producers of wheat. About 5 per cent of the area of crops and pasture was under wheat in 1932 in England

[1] Coppock (1971), 23.

and Wales and this rose by over one-third to 7 per cent by 1934 while all other grain crops declined in area. More important for farmers in the north and west was the direct subsidy for fat cattle, other than cows, authorised in 1934. This was introduced because, owing to its agreements with Argentina and the Dominions, the United Kingdom government had found it impracticable to employ quantitative restrictions sufficient to raise the price of fat cattle at home. This subsidy resulted in modification of the practice of sending store cattle from hill and marginal areas to be fattened in the richer regions. It was more profitable to fatten locally wherever possible. Thus, the 1933 exports of fat cattle from Northern Ireland to Great Britain were more than doubled in 1935 and remained at about this level until 1939.

Subsidies and grants were greatly extended in the years immediately prior to and during the war in order to stimulate home agriculture. We are not concerned with the temporary distortion of the pattern of land use that followed, though this is of interest in showing the extent of the change that can be wrought by financial measures, backed by administrative orders and controls, in an emergency. Let it suffice to note that the area under the plough in England and Wales expanded from about 3·6 million hectares in 1939 to almost 6 million hectares in 1944.

Between 1945 and 1970, assured markets and guaranteed prices under national marketing schemes provided market conditions such as the farmer could hardly have dreamed of in the 'thirties. The Agriculture Act of 1947 was not for nothing called 'The Farmers' Charter'. Assured markets, guaranteed prices, or deficiency payments related to standard prices, were maintained for fat cattle, fat sheep and fat pigs, liquid milk, wool, eggs, potatoes and cereals. Other important products, such as store cattle and store sheep, received good prices because the end product had an assured market.

Entry into the E.E.C. in January 1973 required further extensive adjustments to bring British support for farming into line with the continental practice of restricting imports by quotas imposing levies on permitted imports and using the funds accumulated under the Common Agricultural Policy to subsidise farmers in numerous ways, while also keeping up prices through buying into intervention and storage. Trends already evident —

in farm amalgamation, reduction of labour and changes in farming practice — have continued and, in some cases, have been accentuated by these changes. British farmers, aware of their advantages conferred by a higher state of technological development and generally greater efficiency than their continental neighbours, expected to profit by the change but found varied effects on their fortunes. Arable farmers benefited by increased prices for grains, some livestock farmers gained through opportunities to export more produce, but dairy farmers found themselves exposed to new competition from subsidised exports from Europe and all livestock farmers suffered from rapidly rising prices for feedstuffs.

These changes, superimposed on pre-existing rises in costs which encouraged specialisation and reduction in labour costs, resulted in some major changes being evident in the land use of Britain by 1975 compared with 1970. The number of farms in dairying fell from 110,000 to 81,000 but the average herd rose from 30 to 40 cows. Pig rearing units fell from 66,000 to 35,000, and egg producers from 132,000 to 84,000. Farm employment fell from 425,000 to 375,000. All these trends had, however, been in existence for some years previously and total production of most commodities continued to rise. Important exceptions were butter and some other dairy products, pig-meats, eggs, potatoes, apples and pears, all of which showed marked declines in total output between 1970 and 1975, resulting in the necessity for increased imports and consequent inflationary pressures.

Capital Investment

An important objective of financial aid for the agricultural industry has been the preparation of farms for long-term efficiency by encouraging investment in the land. Thus, when it appeared likely that another world war would force Britain to rely to a much increased extent on home-produced food, the government, from 1937 onward, encouraged land improvement by subsidies for liming, fertilising and bringing old grassland under the plough.

Schemes for comprehensive improvement of farm units were subsequently introduced and these have provided remarkable opportunities for farmers to have their properties overhauled with substantial help from public funds. One of the first such schemes was directed to the improvement of hill farms. The plight of many farmers of hill and marginal land, following decades of low prices, and less favourable conditions even during the war than applied to many other types of farming, resulted in legislation in 1946 which provided for grants of up to 50 per cent of the cost for improvements to land, buildings and equipment of hill farms.

A decade later the principle of assisting comprehensive improvement was extended to all types of farm. The Farm Improvement Scheme of 1957 made grants for improvements to land and buildings, work on farm roads, fencing and installation of electric light and power, and towards the amalgamation of uneconomic units.

Most of the earlier schemes were consolidated in 1971 in the Farm Capital Grant Scheme with similar objectives, but when the United Kingdom joined the E.E.C. it became necessary to adopt the E.E.C. Directive on Farm Modernisation. This necessitated a system of selective aids for the development of farms to the point where the income from the farm for each labour unit employed on it would at least equal the national average income of workers employed in non-agricultural occupations. To try to meet this extremely difficult objective the U.K. introduced the Farm and Horticulture Scheme with a standard rate of grant of 25 per cent for most improvements and 60 per cent for field drainage.

The increased capital investment in British farming in recent years has permitted many changes in type of farming and land use pattern. Thus, dairying, with the high standards of hygiene demanded by government regulations, has been practicable only on farms which could be equipped with buildings, water supply and cattle of approved quality. Farms which have been able to invest in modern machinery have been able to adjust their farming more readily to suit other changing conditions, such as favourable cereal prices, especially for barley, minimising hindrance from lack of labour. Mechanisation has, indeed, been a factor in attracting good quality labour to large, well-capitalised farms. Investment in new houses for farm workers has produced similar results on farms which could sustain such investment. Field drainage, which was a feature of seventeenth–nineteenth-century improvements, but then received relatively little attention until subsidies were introduced in 1940, has been the subject of much increased effort since 1952, with a further dramatic upsurge in area treated after 1965. Over 30 per cent of the arable area of Essex was tile-drained between 1952 and 1971.[1]

In addition to development schemes which apply throughout the United Kingdom, special help has been made available for farmers in areas of special difficulty. Of these the most important in the extent of the region involved is the aid given to crofters in the west of Scotland. A crofter enjoys a special form of tenure, giving him security in respect both of his enclosed land and of his share of the township grazings. This situation derives from the Crofters Act, 1886, which reversed the previous utterly insecure position of the crofter. With the adoption by the E.E.C. in 1975 of a Directive on Less Favoured Farming Areas, minor adjustments had to be made to subsidies and grants paid to crofters. This directive replaced hill sheep and hill cattle subsidies by compensatory allowances of broadly similar effect.

Farming in Britain today must therefore be seen as a response to an extremely complicated set of factors. The range of crops which could be grown in the physical conditions obtaining is narrowed by the availability of imports from abroad, but the Common Market offers a high degree of protection and, together, the E.E.C. and the U.K. government provide a wide range of financial aids to capital investment on farms.

Regional Patterns of Land Use

The complexity of explanations of the land use pattern is evident. Nevertheless, the broad trends follow the indications given at the beginning of this chapter. Thus, in northern and western counties of Ireland, where grass grows freely and arable crops suffer from the damp and windy climate, for many years over 90 per cent of the improved land has been in grass. During the 1970s the proportion of the crop and pasture land actually cropped fell below 1 per cent in Counties Fermanagh in the north and Leitrim in the south. More easterly counties in Ireland also have a high proportion in grass though conditions of soil coupled with reasonable demand for arable crops and the somewhat more favourable climate combine to produce locally more attention to crops. Thus, whereas in the mid-1970s Co. Down and Co. Londonderry had 85–86 per cent of their improved land in grass,

[1] Green (1976), 14–15.

the comparable figure was about 60 per cent in the southern county of Wexford and 70 per cent in Co. Dublin. In these counties soil types are particularly suitable for potatoes, vegetables and grain and, though demand fluctuates, it is usually economically attractive to have a portion of the land in cash crops.

In Great Britain there is a similar broad change of land use from west to east. Not only are much higher proportions of improved land used for arable farming in the east but it is here that the arable acreage is most constant, whereas in the west and north there have been marked fluctuations.[1] Thus, of the improved land of Norfolk, between 70 and 87 per cent has been under arable in each year since 1870, whereas Oxford shows fluctuation between 71 per cent and 35 per cent and Carmarthenshire between 45 per cent and 4 per cent. In mountainous areas, however, where improved land is severely restricted, a high proportion of cultivable land may be under the plough. Thus, Ross and Cromarty, with only 7 per cent of the total area classed as crops and grass (93 per cent rough grazings and deer forests) about 40 per cent of this improved land has been in tillage.

The balance between pastoral and arable use of the land is only one aspect of land use. Arable rotations usually involve periods in temporary grass as well as crops. When tilled, the arable land may be used to produce cash crops for sale or for feeding to livestock on the farm. If the former, the crops may be for human consumption, for industrial use or for feeding to animals. For some crops, such as wheat, the outlets are well defined. Others, such as potatoes, may be used for any of the above outlets, but the objective is normally to sell most of the crop for human consumption, keeping some for the family on the farm. Chats (undersized potatoes) and others which are unmarketable are, however, fed to livestock. Industrial outlets, such as potato crisp factories, exercise an important demand locally, and in some seasons, over large areas. Farms near areas of large consumption, such as the south of England, find less difficulty in disposing of their produce than do areas further away, even with a marketing scheme applicable to all areas of the country. Again, transport costs lower the actual returns to farmers in the peripheral areas although a guaranteed marketing scheme ensures that the farmer will not be left with produce on his hands.

[1] Best and Coppock (1962), esp. 78–82.

Geographical advantages are, then, still important, even under guaranteed marketing schemes and subsidised agriculture. Returns to the farmer will still be highest in general where the alternatives are greatest, because the farmer can there best adapt to changing conditions. Although it may take a few years to change over from say, wheat to beef, or mutton to barley and pigs, it is an advantage to be able to make this change if circumstances in marketing and wider economic conditions make this desirable. This is probably as important an advantage of farmers in the south and east of England as its favourable climate, when the region is compared with the north and west.

In the fertile parts of East Anglia almost all of the farming practices of the British lowlands are possible. Wheat may be grown in rotation with roots and sugar beet, providing cash crops and some stock-feed. This, with some imported feedstuffs, is used for fattening livestock, especially in order to put meat on the market while less favoured regions are waiting for the spring grass. In general, arable crops pay better than grass, so grass is reduced to one year, say, in the rotation. In the midlands of England, where rainfall is heavier, sunshine and warmth less, grass grows better than in East Anglia, but the alternatives in land use are fewer, because cereals and root crops fare less well. These deficiencies become more marked further west and north until one reaches the areas of markedly limited possibilities. The east of Scotland is marginal for wheat, but good grain crops are possible and the cool climate confers a degree of freedom from pests. Livestock are fattened to a considerable extent on the produce of the arable land, feeding in enclosed courts which protect the animals from the rigours of the climate. Among cash crops potatoes are important, especially seed potatoes, again because of freedom from pests and disease.

In the west of Britain including Wales and most of Ireland, as already noted, the high P/E ratio makes cereals definitely risky. There exist some advantages compared with the south-east, such as lower land values, but these do not offset substantially the inferiority of these regions of few alternatives.

These are the broad regional trends, established in the economic conditions of a long period during which British agriculture has faced competition from specialised agricultural regions overseas. In general, agriculture in the British Isles has adapted itself to supply the British market with high quality meat and other products for which there is a degree of natural protection accorded by distance and seasonal variations in supply, such as vegetables. The growth of the large urban communities has facilitated specialisation in the supply of milk and vegetables even where physical conditions are not especially favourable. Thus, milk production is protected by difficulties of transport over long distances. London may draw milk from almost any part of England, and Kintyre may help supply the Scottish lowlands, but Ireland lies outside the area of supply to Great Britain with present slowness and costs of sea transport. In Northern Ireland for decades only 25–30 per cent of the milk produced has been consumed as liquid milk and the rest used for manufacture, whereas in Great Britain 60–70 per cent has been sold as liquid milk. In the inter-war years, this led to Northern Ireland having a different system for payment to milk producers. It was not until after the war that Northern Ireland adopted the British practice of paying farmers a flat rate irrespective of whether their milk was actually used on the high-price liquid market, or for low-price manufacture. Thus, British farmers have had almost a complete monopoly of the home liquid milk market.

In vegetable production, the natural protection of the market is less, but transport of fresh vegetables must be rapid if they are to be marketable as fresh produce. Rarely can imported vegetables command the same price as locally grown ones, so there is economic justification for the devotion of land close to cities to the growing of vegetables. Since horticulture is an intensive form of husbandry, growers are able to resist urban expansion more effectively than can farmers with less intensive systems. Even so, the disappearance of the best market-gardening lands of the London region under the sprawl of the suburbs is well known as a classic problem in town and country planning.

Although improvements in transport have reduced the dependence of urban areas on the vegetables and milk of neighbouring areas, the zoning of farming land around cities is still to be found, demonstrating the overriding of physical conditions by the economic advantages of intensive forms of agriculture. We may not be able to distinguish the concentric circles of von Thünen or the idealised regions of Jonasson (discussed in Chapter 9), but the overlay of local economic power on the broader regional patterns dictated by climate and soil is undeniable.

A great deal of information on the pattern of farming in England and Wales as it was in the late 1960s can be obtained from the Types of Farm maps prepared by the Agricultural Development and Advisory Service, using parish statistics of 1968. Farms were classified into nine types by dominant enterprise[1] and five size classes and the maps are published in sections on a scale of 1:250,000. An alternative approach by Coppock showing combinations of farm enterprises is described in Chapter 9.

A Sheep, Dairying and Forestry Region — Kintyre

The peninsula of Kintyre is part of the major region of the Highlands of Scotland, but is distinctive in that the terrain is less mountainous, the climate milder and the economy different from that of most parts of the Highlands. The individual farm is generally large enough to support a family comfortably. Kintyre is not a crofting region like the western seaboard further north,[2] there being only a handful of smallholdings in part-time cultivation. Some of the farms are part of large estates, but there has been a gradual transfer of ownership through sales of farms to the former tenants. Ownership of land by the farmers rose from 14 per cent in 1913 to 26 per cent in 1956, and in 1975, 72 per cent of all farm land was owned by the occupiers.

The height of much of the land is over 300 m, and the rugged and exposed nature of most of it necessitates hill farming practices. Here are farms of up to 3000 hectares, carrying more than 3000 sheep of the hardy Scottish Blackface breed. Some cattle are carried also on most of these farms, including the long-horned West Highland breed and the equally hardy Galloway. The number of cattle carried must be limited to those that can be supported on the winter fodder available, and there is little land suitable for hay or silage on these farms. A typical farm of 2800 hectares has only 22 hectares classed as arable with a further 8 hectares reseeded to give fairly productive grassland. A herd of 20–30 breeding cows is maintained. The main income of

[1] The classification is described in *Farm Classification in England and Wales 1963*, H.M.S.O., 1965.
[2] Darling (ed.) (1955).

the farm, however, is derived from the 2000 ewes that find their food on the hills, winter as well as summer.

Hill farming represents one of the more specialised types of farming encountered in the British Isles, and, where fodder cannot be provided economically for cattle, farms carry sheep alone. But farms dependent on the sale of virtually only wool and stock — mostly lambs — are economically vulnerable and need to be large to be sound. Wherever possible, some other enterprise should be practised. In Scotland it is considered that a single family hill farm should, to be economic with the low level of profit per head of stock, carry at least 1000 ewes and a herd of cattle. On most hill farms the only practicable cattle are hardy beef breeds, because of limitations of fodder, the lack of fairly level fields suitable for dairy cattle and the cost of transport of milk. Kintyre, however, is more favoured in this respect, and many of the farms that include low ground as well as hill land combine dairying with hill sheep farming. This rare combination comprises a type of farming found elsewhere in Scotland only in the Southern Uplands. A medium-sized Kintyre example of this type of farming has 41 hectares of arable land, supporting about 40 dairy cows, and 170 hectares of hill land on which about 130 ewes are kept.

Near the southern end of the peninsula is the Laggan of Kintyre, an area of almost flat alluvial land. Here, fine pastures have been created and dairying is the principal enterprise with supplementary fattening of sheep and raising of arable crops.

A Region of Small, Mixed Farms — the Lough Neagh Lowlands

Ireland is a country of small farms, owner-occupied (as explained in Chapter 3) and engaged mainly in livestock enterprises, with crops locally important. One of the most mixed regions agriculturally is the basin of Lough Neagh. The Tertiary basalts which were here warped downward, leading ultimately to the formation of the largest inland lake in the British Isles, are covered by glacial and post-glacial drifts. These range from clays to sands and peats. Areas of bog, partly reclaimed, lie between drumlins and kames. Within the basin, many minor land-use regions have been discerned[1] and may be seen to be related to differences of relief, soil and the history of human occupation,

[1] Symons (ed.) (1963), Part IV.

but the larger region provides a convenient unit for the illustration of the mixed farming found here. The farms are organised in small units and devoted principally to livestock, but many variations occur in which enterprises tend to be numerous. Thus, a farm may have five or six cows in milk, twice that number of dry cattle, 30 sheep, some pigs and poultry, a few acres in potatoes for a cash crop as well as domestic use, and oats or barley, some of which may also form a cash crop. Most of the farm is usually in grass, some of which may be harvested for seed as another cash crop. Such a farm is typically of 12–16 hectares and some land may be added temporarily by conacre lease (described on page 66 in Chapter 3).

At the southern end of Lough Neagh most farms have some land devoted to fruit growing. Small fruits are commonly found on the drier ridges among the fen peatlands close to the lake, while on the drumlins stretching away deep into County Armagh, apples are the main crop.

Also close to the lake, especially on the County Tyrone shores, are areas of very small and poor holdings which hardly merit the term 'farm'. Many are of only 2–4 hectares, and much of the work in the fields is still done with the spade. Donkeys are commonly seen here — the only region of Northern Ireland, apart from some areas in Fermanagh, of which this is still true.

The Lough Neagh basin is in the centre of Northern Ireland, with easy access to the Belfast market, and good communications facilitating export to Great Britain of surplus produce. Although not first class,[1] the land is generally of reasonable fertility and, when well managed and particularly when drainage is adequate, can yield good returns in the uses to which it is generally put, as described above. Where poverty exists, it is generally because the holding of land is excessively small for the conditions obtaining. Amalgamation of holdings is gradually reducing the number of such very small units, but many of the resulting farms are still not large enough to be satisfactory economic units. This, indeed, is true of large areas in Ireland.

An Arable Region, Predominantly Grain Growing — Cambridgeshire

Cambridgeshire is selected to illustrate one of the regional variations in types of farming which occur within the mainly

[1] See Chapter 10 (p. 237) for relevant land classification.

arable area of eastern England. The county is generally recognised as lying within a grain growing belt associated with the
occurrence of fairly large farms in Cambridgeshire, Suffolk and
adjacent counties. Rarely is grain growing so dominant as to
exclude all other enterprises, but in this central belt there is a
higher proportion of farms on which it is clearly predominant
than in the adjacent counties where roots, both for stock-feed and
cash (particularly sugar beet) rival or exceed grain in importance.
The emphasis on grain is found on a variety of soils, including
loamy, chalky and clay types, with wheat favoured on clays and
barley on chalk.

Grain farms, such as occur in all parts of the eastern counties
except the Fens, have an average area in cereals equivalent to 70
per cent of the *arable* area and average 75 hectares in total size.[1] A
minor livestock enterprise, usually beef cattle is commonly found
on these farms.

The combination of an important area of roots with grain is
found on farms of similar size (average total size 75 hectares, but
with a high proportion in the medium-size category of 40–120
hectares). In this category also, nearly half of the farms have a
livestock enterprise, usually beef cattle. Output of cash roots
becomes more important than grain on smaller farms, of which
there are examples in north Cambridgeshire, where they are
much more numerous on the fringe of the Fen country. Sugar
beet is the typical cash root crop in both types of this combination,
with potatoes also favoured, mainly on the medium and large
farms.

This region fits the general assumption that arable farming is
associated in Britain with the areas of lowest rainfall, avoiding the
low-lying areas of medium and heavy soils on which dairying is
favoured. Grain growing occupies a predominant place on larger
farms, which can obtain a satisfactory income from it, in spite of
its comparatively extensive nature. It is not, however, usually the
only enterprise and the maintenance of livestock qualifies the
farms for inclusion in the broad category of mixed farms
characteristic of the United Kingdom, but with the important
qualification that each farm tends to have fewer enterprises than
would have been the case in earlier decades, a tendency clearly
reflected also in other areas.

[1] Jackson, Barnard and Sturrock (1963).

New Zealand

The islands of New Zealand lie in temperate latitudes and the observed means for temperature and rainfall place most of New Zealand in the same category as Great Britain in the Köppen classification — Cfb. Averages here tend to mislead, for New Zealand enjoys higher daytime temperatures than Britain, hours of sunshine are longer and snow is rarely seen in the lowlands of New Zealand. The north of New Zealand is warm temperate rather than cool temperate, as may be seen in the successful culture of the vine and citrus fruits. Throughout the islands, however, the climate is humid, except for some of the interior montane basins of the South Island, notably Central Otago.

The combination of moderate temperatures and fairly high humidity gives nearly ideal conditions for growth of grass, and it is on the high-yielding pastures of sown grasses and clovers, maintained with scientific management and heavy fertiliser applications, that New Zealand's prosperity has been based. Only on a narrow belt of the west coast are lowland pastures endangered by excessive rain, but drought can be a serious threat in most other regions. Where grasslands existed at the time of European colonisation, they were composed of native species, mainly of tussock habit, of low grazing value. These had to be replaced by English grasses, and the nitrogen cycle based on clovers had to be harnessed before potential fertility of the lowlands could be developed.[1]

Hence, British settlers who came to New Zealand via Australia found conditions more like the homeland than they had found in Australia, but those who had come direct would have noticed the differences more sharply.

The European colonisation of New Zealand did not begin until 1840. The Southern Alps had been sighted in 1642 by Abel Tasman, and Cook had made landings and surveyed the coasts between 1769 and 1777 but the only shore stations successfully maintained before 1840 were for sealing and whaling. Consequently, by the time permanent settlement was undertaken the colonists could draw on the knowledge of the improvers of

[1] Curry (1962, 1963) examines the relationships between climate, grass growth and livestock feed requirements on New Zealand farms. For a brief description of the grassland of New Zealand, see Sears (1962).

agriculture in Britain, and the experience gained in developing agricultural land in America and Australia.

Initially, cultivation was for local requirements, there being no accessible overseas markets on which crops could be sold. Sheep, however, brought in the first instance from Australia, provided an exportable product, wool, and during the 'sixties, there developed a substantial trade in wool. Meanwhile, the four main centres and a number of lesser towns had grown sufficiently to provide an urban population large enough to offer outlets for a variety of crops. But it was only with the introduction of refrigeration in 1882 that it became possible to realise fully the potential productivity inherent in the land and climate of New Zealand. Refrigeration made it feasible to expand the pastoral industries to supply the great market in Britain with meat and dairy products. Whereas, compared with Australia, New Zealand had played only a small part in the wool trade, the climate was far more suitable than Australia's for dairy produce. For meat production also, conditions were favourable and New Zealand farmers began the changeover from the Merino, which produced fine wool but a poor carcase, to types of sheep which were more suitable for meat production. As the English market reflected the growing preference for small joints of meat, so New Zealand production settled into concentration on fat lamb production.

Great Britain alone in all the world offered open and remunerative markets for the agricultural products of temperate lands. Nowhere else was there the great imbalance between food requirements and production, backed by the purchasing power conferred by massive exports of manufactured goods and services. New Zealand had the advantage of being a British colony, which was important for trade contacts, but, even had it not been, it could have competed on the British market as did the United States, Argentina and Denmark.

New Zealand suffered from one major drawback — distance from Britain. At 22,500 km, via Australia, it was about as far as it could be from its market. The opening of the Panama Canal in 1914 reduced the distance to 19,300 km. Although transport has been speeded up, cargo liners on the New Zealand run still normally take four to five weeks on a direct voyage. Furthermore, before the voyage commences, loading of cargo involves a laborious and time-consuming tour of small ports around the

New Zealand coast, though the number of ports visited has been reduced in recent years in the search for economy of operation. The consequence of this long haul is not only that transport costs are high but that the producer must be prepared for considerable fluctuations in prices before his produce is finally marketed, months after it left the farm or factory. The normal difficulty of any farmer in trying to anticipate conditions of surplus and shortage is made considerably more difficult. The New Zealand farmer has also to face the fact that his produce is frozen and will never command the highest prices, which are reserved for the fresh commodities marketed by those nearer at hand.

To meet these disadvantages it has been essential to organise production to exploit all the conditions that favour the new Zealander, and to refine the system progressively. By the 1920s it had been proved that in New Zealand it was practicable to base stock farming almost entirely on grass and clover, maintained with heavy dressings of superphosphate. Grass growth is at a standstill for only short periods — longest, of course, in the south of South Island, where supplementary fodder is needed — and little housing of stock is required. But these factors would not reduce costs sufficiently for New Zealand to compete if farms were small and diversified like the majority of European farms. The New Zealander has had to learn to specialise and to accept the risks inherent in specialisation — risks which are all the greater when the specialisation is directed at an overseas market. Specialisation in search of low costs results in there rarely being more than two enterprises on a farm, of which one is usually pre-eminent. The majority of farms can be classed as dairying or livestock-producing, or a combination of these two. Sheep are combined with cattle for maximum pasture utilisation and control of weeds.

Provision of winter fodder is organised so as to minimise labour requirements and cost. Hay-making has long been mechanised, though silage is widely preferred to hay. In the south, turnips, rape, kale and oats are important fodder, so arable land is seen in conjunction with livestock. Turnips illustrate the economy of man-power. They are not lifted for feeding to stock, the sheep being merely turned into them to eat their way steadily through the crop, controlled for thorough utilisation by temporary fences. The number of hectares and of animals managed per worker is large by world standards, excluding collective farms. The average dairy herd has about 110 cows. Mechanisation of the milking came at an early date and almost all dairy cows are now machine milked. Similarly, with several thousand sheep to be shorn on a farm, commonly in a few days by itinerant shearers, the woolshed also attracted mechanisation and for many years most sheep have been machine-shorn. Employment on New Zealand farms is limited not only because wages are high but because labour is perennially scarce. Shortage of labour has always been a handicap in New Zealand, following inevitably from a policy of restricted immigration and the distance of the home country, the main source of settlers, and the labour force has been steadily diminished by migration to urban occupations.

The typical New Zealand farmer assesses his livelihood in pounds of butterfat available for the factory, number of lambs for the freezing works, and weight of fleeces for the woolbroker. His aim is to carry as many livestock as he can feed. Number of livestock per man is the yardstick of profitability, but good quality land is limited in supply, so number of stock is also high per hectare in the good quality lands, and often higher than it ought to be on poor, mountainous land. The farms of New Zealand are large by European standards. In 1975, 46 per cent of all farm holdings exceeded 100 hectares in size and 7 per cent exceeded 800 hectares. The latter group included the high country sheep stations, which ranged up to 40,000 hectares. At the other extreme, there were a good many rather small farms, 6500 (10 per cent) being of between 20 and 40 hectares. This, however, was only about one-half as many as a decade earlier.

The need for large farms, permitting large numbers of stock to be carried without detriment to land or animal health is seen when prices for produce are compared.

In Great Britain prices rose steadily through the 1960s and accelerated in the 1970s. The average price for fat sheep and lambs in England and Wales rose from 16·2 pence per pound dressed carcase weight in 1965 to 24·8 pence in 1972. In the same years New Zealand's average export schedule price fell from 21·5 cents per pound to 15·6 cents and at times was below 13 cents.[1] Since, in general, the New Zealand cent approximates to one halfpenny it will be seen that New Zealand farmers were

[1] Commonwealth Secretariat (1973), *Meat, a review*, London, 53.

producing fat lamb at prices that would have been considered quite uneconomic in Britain — or elsewhere in Europe. Since 1972 the position of the New Zealand farmer has worsened as, although prices have continued to rise in Britain, a levy was introduced on imported mutton and lamb, followed by the first phase of the E.E.C. tariff introduced in 1974 at a rate of 8 per cent, rising to 20 per cent in January 1977. The cash returned to sheep farmers in New Zealand has, therefore, risen only slightly while the costs of their imports of machinery and other farm equipment and fuel have risen rapidly with inflation in Britain and other industrial countries and the increases in oil prices. New Zealand dairy farmers have suffered even more because there are fewer alternative markets, but they can sell their products on the British market a little cheaper than the home-produced, unfrozen product. They could also easily undersell the home producers in other European countries, North America, and, in dairy produce, Australia. Quota restrictions, or complete prohibition, however, reinforce tariffs in keeping out the genuinely low-cost produce from New Zealand, or in admitting only as much as is needed to balance supply with demand at the prices acceptable to the home farmer. To reduce its dependence on the British market, New Zealand has developed new markets, particularly in Asia. There, of course, the demand for food is immense, but the effective demand for New Zealand produce is limited to some extent by the lack of familiarity with temperate foodstuffs and, ultimately more seriously, by the lack of purchasing power.

Clearly, in a country as dependent as New Zealand is on its agricultural produce — which is responsible for about 80 per cent of its overseas earnings — farmers cannot be subsidised nor can prices be kept at artificially high levels. The most that can be done is to even out fluctuations in prices from funds accumulated when trading prices are high, and to make government grants available for certain purposes judged to be important to the nation as a whole, such as soil conservation measures and the improvement of marginal land.

Land Development in New Zealand

In spite of the restricted markets open to New Zealand and the relatively small, though growing, internal market (1975 population 3·1 million) production has been increased markedly in recent years. Until 1955 the annual output of lamb was under 200,000 tonnes, in 1972 it reached 379,000 tonnes.[1] Increases in livestock numbers are achieved partly by more intensive use of land, but land development continues to be, as it has always been, a feature of the New Zealand scene. Problems of allocation and tenure of land have always loomed large in New Zealand's internal politics. As the population has grown and export earnings have had to be increased, Crown land held in reserve by the government has been released for settlement, and large farms and sheep runs have had to be subdivided and worked more intensively.[2] High valuation and taxation of large estates were among measures used to achieve this end. Many estates were converted into small farms for the resettlement of returned soldiers after the First World War, and of these a high proportion were too small to support a family. Learning from this experience, present-day practices produce fewer but larger farms.

The state is the main agent of land development in New Zealand. Many of the properties taken over by the Lands and Survey Department after it became the responsible authority in 1929 were already in grass, and simply required subdivision, and provision of roads, buildings and services. Increasingly, however, development work has been concerned with land previously considered to be uncultivable. Experiments beyond the means of the private individual, and the economies of scale in machinery, materials and labour, have enabled the government to achieve striking results.

Development of quite small blocks of a few thousand hectares may be undertaken, but blocks of 20,000 hectares or more offer disproportionately greater scope, especially if the terrain is difficult and demands a long-term approach. Initial tasks include destruction of scrub and other existing vegetation by crushing, raking and burning and the construction of roads and drains. The land is then ploughed or cultivated with giant discs. Discs have the advantage over ploughs on steep and stony ground. If discs cannot cope with the terrain it may still be improved by oversowing without cultivation. Lime is applied and pasture established. Cattle and sheep can then be brought in, the aim being to stock the land as densely as possible to improve the pastures through balanced grazing and natural nitrogen enrich-

[1] *New Zealand Official Yearbook*, 1975.   [2] Duncan (1962).

ment. During this period houses and fences are built and roads improved.

By this approach about 775,000 hectares had been made available for settlement in over 4300 farms by 1973, and a further 500,000 hectares were under development. A typical year's work includes laying down to grass an area of about 20,000 hectares. Areas are turned over to private farmers ready to support ten or twenty times the grazing pressure which they received when they were being used for extensive grazing before development began. Selection of the farmers to take over the completed property is by ballot with various qualifications required of candidates.

Areas reclaimed in this way are found in all parts of New Zealand.[1] The biggest blocks have been in the central volcanic plateau of the North Island, including the pumice lands, which could be agriculturally developed only after the discovery of the importance of trace elements, the lack of which had prevented livestock from thriving. There are other large blocks, however, from the harsh 'gumlands' (former Kauri forest country, where natural soils are acid and infertile) of North Auckland to cool and humid Southland. After having been rather neglected in the earlier phases of development, when large reserves of Crown land were available in the North Island, Southland has become a region of major importance in land development. By 1974 it was the most important area for this work in the South Island, having 141,000 hectares under development. Areas under development include extensive areas of coastal peat land.

In this venture the Department operates without subsidy from the exchequer and therefore aims for a working profit. Land development costs per hectare, however, vary greatly, and the land most expensive to improve is not necessarily the most valuable. Disposal prices of farms reflect their real worth and hence the state subsidises its less profitable ventures from profitable ones.

The Department of Maori Affairs is also involved in land improvement in the North Island, and private interests have undertaken some large scale development work as well as normal farm improvement schemes.

A feature of agricultural improvement in New Zealand which

[1] Stover (1969) reviews areas, methods and achievements in some detail.

has attracted much attention abroad is the large scale use of aircraft. Although aviation had become an accepted ally of agriculture in many countries, mainly for spraying crops with insecticides, it was New Zealand that developed the application of the light aeroplane to top-dressing and oversowing as a normal aspect of farm operations. It has made possible the improvement of steep and rolling country, particularly the tussock grasslands that have become depleted with fifty to a hundred years of grazing, for which ordinary methods of top-dressing were impracticable. It is also used on all other major types of terrain. The total quantities of fertiliser (mainly superphosphate) and lime applied from the air have risen year by year and generally exceed 900,000 tonnes, treating 2,500,000 hectares. This is about the same area as is top-dressed by surface methods.[1]

Aerial top-dressing, followed by better subdivision, the construction of access tracks and stockwater dams and the introduction of more productive grasses has been described as one of the two major technological revolutions benefiting New Zealand farming in the twentieth century, the other being the revolution in grassland management in the 1920s.[2]

Aircraft are also used for sowing seed, spreading insecticides, fungicides and poison to combat rabbits and weeds, distributing trace elements, dropping fencing materials in remote areas, spotting stock on large runs and many other tasks.

Irrigation is another aid to more intensive use of land which has received increasing attention in recent years in New Zealand. In a number of districts, notably South Canterbury and Central Otago, production is retarded by long dry periods. Irrigation has been developed for stock fattening and fruit growing in particular and, because of the high cost of labour, attention has been given to automatic operation of sluices. Much more land could benefit from irrigation, given the market incentives.

The Regional Pattern[3]

The principal types of farming and their regional distribution appear in Figure 6. Relief is the fundamental factor influencing the distribution of the main types. In his pioneer study which

---

[1] New Zealand, *Agricultural statistics* (annual).
[2] Stephens (1976).
[3] For regional descriptions in more detail, see Cumberland and Fox (1970).

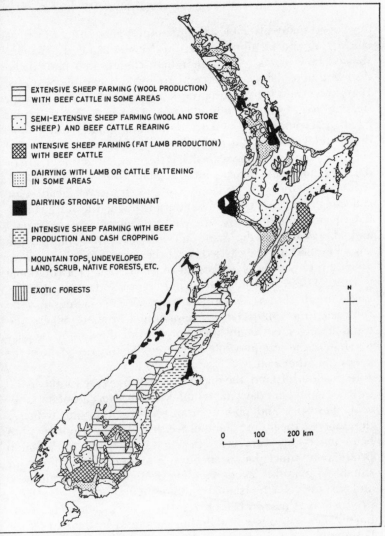

EXTENSIVE SHEEP FARMING (WOOL PRODUCTION)
WITH BEEF CATTLE IN SOME AREAS

SEMI-EXTENSIVE SHEEP FARMING (WOOL AND STORE
SHEEP) AND BEEF CATTLE REARING

INTENSIVE SHEEP FARMING (FAT LAMB PRODUCTION)
WITH BEEF CATTLE

DAIRYING WITH LAMB OR CATTLE FATTENING
IN SOME AREAS

DAIRYING STRONGLY PREDOMINANT

INTENSIVE SHEEP FARMING WITH BEEF
PRODUCTION AND CASH CROPPING

MOUNTAIN TOPS, UNDEVELOPED
LAND, SCRUB, NATIVE FORESTS, ETC.

EXOTIC FORESTS

N

0   100   200 km

Figure 6. Types of farming, New Zealand. Intensification of production has reduced the area devoted to extensive sheep farming which is now important only in the South Island mountain zone, east of the Southern Alps. Much of the low hill country of the North Island is, however, unsuitable for enterprises more intensive than raising store sheep combined with wool production and cattle rearing. The highly productive areas therefore represent only a small proportion of the total area. The great extent of the agriculturally unproductive land in the mountains of the South Island and on the North Island volcanic plateau is evident. The exotic forests are softwood plantations.

Source: Based on maps in *Descriptive Atlas of New Zealand, New Zealand Atlas* (1976) and other sources.

developed many principles of agricultural geography as well as elucidating the farming pattern of New Zealand, Buchanan[1] showed how it was relief rather than climate that limited the distribution of dairy cows as compared with beef cattle or sheep. Dairy cows cannot give high production if they have to cope with steep terrain, even if the altitude is low. Hence, the North Island, having the combination of flat and undulating valley, terrace and downland country with typically over 1200 mm rainfall and yet with a high sunshine record, has the greatest development of dairying, and some 90 per cent of all dairy cows in New Zealand. A further factor is the availability of good transport facilities for collecting milk from the farms and transferring the output of the dairy factories to the ports. The Waikato, Bay of Plenty, Taranaki and Manawatu are the main dairying districts. Typically, two cows per hectare are carried, with the farm, entirely in grass, divided into paddocks of about two hectares each for rotational grazing. These are essentially family farms and extra labour is rarely hired, but seasonal jobs such as hay baling may be contracted out. Many farms are operated by sharemilkers, the usual arrangement being for the owner of the land to be paid one-half of the receipts while the sharemilker (owner of the herd) keeps the other half.

In the South Island, the highest rainfall areas are found on the west coast but here dairying is limited by the shortage of flat land, acid, gley soils and poor communications. Dairying is better developed in the humid plains of Southland, but the tendency has been in recent years for dairying to give way to fat lamb production, which has been more remunerative for farms of suitable size, and is less demanding in labour. Many farms here and in the North Island combine dairying and fat lamb production. A narrow belt of country on the sub-humid Canterbury Plains specialises in dairying, favoured by the market for liquid milk in Christchurch and the facilities for export through the port of Lyttelton.

Dairying is encouraged by the heavier soils and where these give way to lighter textures, and in drier areas without severe decline in fertility, intensive sheep farming is more generally found. Fifteen or more sheep per hectare are carried on light land which responds to top-dressing, with grass and clovers

[1] Buchanan (1935).

supplying almost all feed requirements in the North Island but with arable crops necessary in the South Island to supplement pastures. Irrigation is used in some cases. Fat lambs form the basis of the farm economy, with wool, ewe mutton and beef subsidiary. The larger areas devoted to this kind of farming are the valleys and lower hill country of the east of the North Island, the Canterbury Plains, Central Otago and Southland in the South Island.

On the higher and steeper country, accounting for most of the rest of the occupied land in the North Island, and the foothills and lower ranges of the South Island, more extensive sheep farming is practised. Store sheep for fattening on the low-ground farms and wool production are together the main sources of income on these farms. Conditions vary widely, but between two and five sheep per hectare would be normal, a carrying capacity similar to that of the more productive hill farms of the British Isles.

Until the 1960s cattle rearing was very much a subsidiary enterprise. Expansion of beef cattle herds followed the opening up of the American beef market and on all types of sheep rearing farms and in all the areas of low and medium altitude land the carrying of cattle became common. Many individual farmers, however, preferred to continue to concentrate on sheep with which they were familiar, especially as cattle needed more improvements, such as fencing and water supplies and more winter feed.

On the highest and steepest occupied land, including the ranges of 2000–2500 metres on the eastern side of the Southern Alps, the high country sheep run is virtually the only feasible type of farm. On these properties, extending to many thousands of hectares, hardy sheep are kept for their wool, often with only one sheep to four hectares. Merinos are favoured for the high runs. They climb to the high ridges and eat down to the lower levels, thus coming into easier country before the winter snowfalls, although losses in snow are still heavy. On lower ground some cattle are kept together with crossbred sheep.

New Zealand had, in 1972, 60 million sheep and more than half of its total area is devoted to sheep farming of some kind. Since nearly one-third of the total land area of 27 million hectares is mountainous or otherwise at present unusable, the area in sheep farming (13 million hectares) amounts to about three-quarters of the total agricultural area. From the early days, when the Merinos formed almost the entire national flock, breeds have multiplied to suit the differing conditions of semi-intensive and intensive sheep farming. Flocks on these farms are based on descendents of the English breeds, particularly Romney, Southdown, English Leicester and Border Leicester. To suit South Island conditions the Corriedale was evolved from crossing Merino, Lincoln and English Leicester breeds. In 1973, dairy cattle numbered over 3,150,000, with over two million cows in milk. Beef cattle had, however, been increased rapidly in anticipation of improved markets (which proved disappointing) and numbered nearly 6 million. Dairy cows are predominantly Jersey, favoured because of the high butterfat content of the milk, and Friesian, preferred for town milk supply. Beef breeds are mainly Aberdeen Angus, Hereford and Galloway.

In spite of diversification, in 1973–74 livestock products accounted for 80 per cent of New Zealand's export earnings. Arable crops are grown almost entirely for home consumption, and do not satisfy the domestic market. Wheat was grown for export in the early 1880s, but the specialisation that developed in North America and Australia cut prices below the level at which New Zealand farms could compete, although yields per acre were much heavier in New Zealand, and former grain farms were turned over to livestock production. Today, wheat is an important aspect of farm production only in the South Island, nearly 70 per cent of New Zealand's production being grown in Canterbury, and most of the rest in Otago. Other crops, including oats, barley, potatoes and green fodder crops, and temporary pastures feature in rotation with wheat. Stock fattening is also commonly combined with the arable farming. This is the type of farming shown in Figure 6 as intensive sheep farming with beef production and cash cropping, and is associated, of course, with flat and gently rolling terrain.

Pig and poultry farming, originally developed as sidelines, are now predominantly specialised branches. Pigs, fed on skimmed milk before the introduction of tanker collection of whole milk, are now fattened on meal. Egg production and broiler chicken production tend to be organised in separate, highly specialised units. Access to markets is a major locating factor for these units.

Specialisation in New Zealand Farming

Compared with many farms in Britain and much of the continent of Europe, those of New Zealand are relatively specialised. Apart, however, from holdings devoted to poultry, pigs, fruit and vegetable growing which are similarly specialised in Europe, there is rarely only one significant enterprise on the New Zealand farm, except in dairying and high country wool production. These again, are paralleled in the larger British dairy and hill sheep farms. It is probably true to say that the majority of New Zealand farms have two significant enterprises, such as dairying and fat lamb, or store sheep and cattle, where British farms frequently have three or four. This difference is diminishing as pigs and poultry increasingly are reared on specialised farms in both countries.

The latitudinal extent of the two countries is similar, but this factor induces greater variety in the case of New Zealand because it extends into a warm temperate type of climate, so Mediterranean and sub-tropical fruits, maize and tobacco can be grown commercially. Relief also is more marked, but the extensive sheep farming of the Southern Alps is not more different than hill farming in the British mountains from their low country equivalents. Some crops important in Britain are absent, notably sugar beet. But the outstanding difference is in social and economic organisation. Throughout New Zealand, conditions are comparatively uniform in the range of sizes of farms, the marked predominance of freehold tenure and rural land values, and marketing opportunities. The differences that exist in Britain, which have evolved with the different historical experiences of different regions of Britain, superimposed on differences of climate and relief, have no equivalent in recently-settled New Zealand. Enhancing the uniformity is the orientation of the country to distant export markets, and the need to produce the commodities in which the country's advantages can be most fully realised.

The system thus developed over a century or so is now in jeopardy. Re-orientation of production and marketing was begun as soon as it seemed likely that the United Kingdom would enter the E.E.C. with the consequence that imports from New Zealand would be curtailed in accordance with the E.E.C. policy of restricting the competition faced by European farmers. The lack of alternative markets, however, necessitated New Zealand farmers being prepared to accept minimal returns in order to continue to compete in Britain for as long as they were permitted to do so. In 1976 dairy farmers in New Zealand were receiving about one-half the financial return per tonne of butterfat that E.E.C. farmers were getting. This could be tolerated because of the greater efficiency of production in New Zealand but for the quantitative restrictions which have steadily reduced the amount that Britain could import from New Zealand. European producers sought a further reduction of 20 per cent in the amounts to be permitted between 1978 and 1980 to bring the annual amount down to only 60 per cent of the 1968 figure, with no guarantee of access to the market after 1980. Sheep farmers have also faced reduction in demand consequent on rising prices as the levy on imports into Britain rose in accordance with E.E.C. regulations and the possibility of quantitative restrictions in this case also had to be considered.

At a time when prices for imported and other manufactured goods were rising sharply, the New Zealand farm sector suffered a decline in its share of the national income from 7 per cent in 1973–74 to about 3 per cent in 1974–75.[1] This was accounted for by large falls in incomes on individual farms, especially dairy farms, which would have been even worse but for increased amalgamation of holdings as many farmers moved out of agriculture. Incomes in manufacturing and other occupations seem more attractive though it is impossible to forecast how foreign exchange will be earned to pay for imports on the scale necessary to maintain secondary and tertiary industries if agriculture can no longer do so, since it still accounts for nearly 80 per cent of total overseas earnings. No clearer example can be found of the way in which political decisions, expressed through economic measures, can radically change and perhaps destroy a system of farming superior to any other in the world in price/cost efficiency terms, causing mutual loss of benefits for the populations of the two countries that had developed the original trading partnership.

[1] *ANZ Quarterly Survey*, July 1975, 7.

CHAPTER 6

Plantation Agriculture in Malaysia

The antiquity of agriculture in south-east Asia is well known as a result of archaeological investigation. The view that this region was probably the cradle of the very earliest agriculture, and the home of domestic animals, has been referred to above (Chapter 1). From these early beginnings, however, the people of the region developed their techniques comparatively little, and at the present time subsistence farming in the region is still based on the practices of hundreds and even thousands of years ago. In this chapter we shall consider some of the features of the alien plantation system of agriculture which was introduced into the region by Europeans for their own commercial gain.

Prior to the introduction of plantations, the exports of the region were confined to spices and other luxury items, which, being highly prized by wealthy society in Europe, could withstand the high cost of shipment and long duration of travel by sailing vessel. Indian traders dominated the trade in spices, perfumes, rare woods, alluvial gold and precious stones for over a thousand years before the decline in the thirteenth century of the state of Sri Vijaya, the Sumatran focus of Indian culture in the archipelago. Power became increasingly concentrated in the hands of Muslims, with Arab traders and missionaries challenging the Indian interests. Then the Europeans arrived and began to set up trading posts. Malacca was captured by the Portuguese in 1511 and a Portuguese agent was established in Sumatra in 1512. A hundred years later, supremacy in the region was passing to the Dutch, who were not seriously challenged until the end of the eighteenth century, when the British secured Penang.

Up to this time the Malayan peninsula had been of less commercial importance than the islands, even though the first European base had been at Malacca. But the British saw the potential value of the swampy and almost uninhabited island of

Singapore as the strategic centre for both peninsula and islands, and something at least of the scope for producing tropical commodities in the Malay states. They could not at that time have foreseen that the most notable contribution that Malaysia would make to the twentieth century world economy would be through the production of natural rubber.

By the nineteenth century the commercial exploitation of the tropics by the European nations had of course already been in progress for several hundred years. The plantation system had been found to answer the needs of foreign producers, and had been applied successively to different crops and in numerous tropical and sub-tropical environments. The essence of the plantation system was the acquisition by Europeans of sufficient areas of suitable land to make economically attractive units, on which were established European managers and assistants to provide commercial organisation and technical direction. Plantations later established by companies were similarly organised. A number of plantation units were required to justify port and other commercial facilities essential for conveying the products to the centres of demand in Europe and, later, in North America. Manual labour had to be cheap so that the product could be sold at a price which would ensure a large and growing volume of demand. Workers had also to be capable of toiling long hours in tropical heat. In many cases the indigenous population was either too scanty to form the basis of a labour force, or was not readily amenable to discipline. Hence the wholly reprehensible system of slavery and later the objectionable practice of indenturing were introduced to deal with the labour problem. The plantations of the Americas flourished on these supplies of European and North American capital and forced labour, and most of the profits were repatriated to the sources of the capital.

A number of crops were tried with varying degrees of success on plantations in Malaysia in the nineteenth century. Cloves, nutmegs and pepper were followed by more extensive plantings of sugar cane and coffee, each of which reached about 20,000 hectares at its peak period. Plant diseases and competition from other producing regions led to the decline of all of these. At the turn of the century, however, rubber was being introduced and inter-planted with coffee and sugar cane.

The value of rubber had been recognised before the middle of the nineteenth century with Mackintosh's utilisation of it for waterproofing in 1823, and, much more important, the success of the vulcanising process. This, in 1842, heralded the era of road transport, though pneumatic tyres for motor cars did not go into production till decades later. The best source was found to be *Hevea brasiliensis*, which will yield a large flow of latex over many years, in contrast, for example, to the Guayule shrub (*Parthenium argentatum*), from which the natural rubber can be extracted only by destroying and crushing the plant. The scattered occurrence of *Hevea* in its natural home, the Amazon forest, and the difficulties encountered with labour in this region led to the establishment of the tree in south-east Asia, which was physically suitable. Although the indigenous Malays were relatively few in number and did not prove to be sufficiently interested in wage employment, the labour problem in Malaya was solved by recruitment in south India, where there was a large surplus population. Chinese immigrants also provided a source of labour, though they were much less important as direct plantation labour.

Humid tropical conditions are essential for the commercial exploitation of *Hevea*. Even in equatorial areas most rapid growth is confined to the lowlands below 200 metres (650 feet). For every increment in altitude of this magnitude trees require three to six months longer to reach the size at which tapping is commenced, and 700 metres (2300 feet) is a practical upper limit.[1] Rainfall of between 1800 and 3500 mm per year is preferred, with no month normally much below 75 mm. At the same time, wet days interfere with tapping, so that it is best if the rainfall is concentrated on not more than about 150 days.[2]

Soils need to be deep, permeable and fertile for the best growth. The fertile volcanic and alluvial soils in south-east Asia give excellent results but the widespread need to use these best soils for food crops means that lateritic soils have to be used for rubber. To maintain good physical conditions, attention has to be paid always to drainage, fertilising and control of erosion, and irrigation may sometimes be used with advantage. The soil is also generally protected by a cover crop between the trees. Soils within

[1] Ochse and others (1961), 950.
[2] For the effects of physical conditions on rubber production in Malaya see Wycherley (1963).

the range pH 4·0 to 8·0 are usable, but the best results are attained with pH held between 5·0 and 6·0.

The tree secretes its latex in tubes in the soft tissues of the inner bark close to the wood. There is no latex in either the wood or the outer bark. The latex is tapped by paring away a thin sliver of bark which severs the latex vessels. After a few hours coagulation takes place at the ends of the vessels so that the flow ceases. Tapping is repeated every second or third day, rubber and bark regenerating naturally. Regular tapping may be maintained over a lifetime of some 30 years but skilful and properly regulated tapping is essential.

Economics of Rubber Production

It is evident that scientific management is required for good results and an appreciable amount of capital is involved. Although the trees grow fast there is a lag of some five years before the first returns are obtained from young trees. Capital is invested not only in trees but in factories for processing the latex (less than one-third of which by weight is rubber), housing for the labour force, roads, vehicles and other equipment. Capital was readily attracted early in the present century, demand having forced up prices for rubber, and governments encouraged companies by allocating land and building roads and railways. Migration of workers was also encouraged, and Singapore developed as a great processing and commercial centre. The volume of shipping between Singapore and Ceylon and the consuming countries in Europe and North America, together with comparatively good internal communications, enabled the plantation product from this region to compete easily with the output from South America.

However, as with other primary products, supply and demand were never nicely balanced for long, so that prices fluctuated continuously. After the First World War prices fell to a very low level, as in other forms of agricultural produce, and to meet this situation the Stevenson Valorisation Scheme was introduced to raise prices by restricting output. This could succeed only if there was full co-operation between producers, but this was not achieved. In particular, the Dutch increased their plantings and production. The situation was worsened by the world economic depression and the price fell until in 1931 it was only $2\frac{1}{2}$d. per pound ($2\frac{1}{4}$p per kg) compared with 12s. 6d. (65p) per pound reached twenty years earlier. This situation could not continue and resulted in 1934 in the International Rubber Regulation Agreement. With excess supplies being withheld from markets some stability of prices was achieved until the Second World War.

Meanwhile, synthetic rubber production had been developed, arising out of Germany's needs in the First World War. It represented only 8 per cent of world consumption of rubber in 1940, but there was a rapid increase in its production during the Second World War. Consequently, the restoration in 1945 of the rubber plantations, after the wartime destruction and severance from markets, had to face new conditions in competition. The range of products in which natural rubber still held distinct advantages over the synthetic product was much reduced, and there was surplus capacity in synthetic plants. Nevertheless, the natural product has itself been developed to be more competitive, and has succeeded in again proving economically feasible.

The world depression which followed the spectacular increases in oil prices forced through by the Middle East producers in 1973 resulted in a drastic reduction in demand for rubber with a price fall from 50 pence per kg to 23 pence in ten months, but gradual recovery took the price back to over 50 pence by mid-1976. This price is, however, moderate in relation to rising costs and in 1975 rubber's contribution to export earnings, at 18·3 per cent of the total, was much lower than a decade earlier. However, rubber still occupied 52 per cent of the cropped area of Malaysia and 53 per cent of the cropped land of the peninsula (Figure 7) so it remains extremely important in the country's economy.

In the present circumstances, estates of less than 1000 hectares are no longer economically attractive to commercial companies, and most of the foreign companies still operating control several such estates.

The profits of estates of this kind have to be judged in relation to the need to find dividends for overseas investors on a scale sufficient to offset the political and economic risks inherent in investment in rubber production. Smaller estates (less than 200 hectares) are acceptable to local owners who do not have this problem. In addition there are a large number of smallholdings locally owned and worked. In many cases these have only a few hectares or so of rubber, integrated with other crops.

For all producers, but most of all for the large commercial estates, the problem today is to obtain the highest possible yield per tapper, because tapping is the largest single item in production costs. Increasing the frequency of tapping or the number of trees per hectare may result in higher yield per

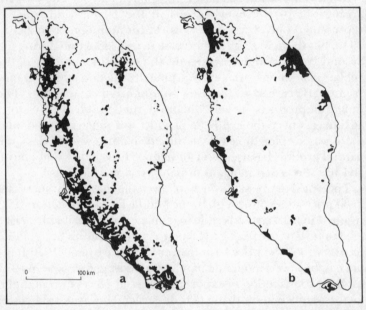

Figure 7. Distribution of land used for production of rubber (left) and other crops (right) in peninsular Malaysia. The remaining land is forest, jungle, or is undeveloped except for the small area of urban and industrial development.
Source: Malaya Land Utilisation Map (1 : 760,320, 1953), reproduced by permission of the Directorate of National Mapping, Malaysia. Copyright reserved.

hectare, but lower yield per tapper, and is accordingly undesirable. Efficiency is therefore judged by the standards common in commercial farming in countries where labour is short, rather than by the criterion of yield per hectare normal for food crops in tropical Asia. However, attention to plant breeding and scientific management have made possible greatly increased yields in recent years. New high-yielding strains produce some 2000 kg/ha compared with about 500 kg/ha normal for older rubber. Extensive replanting with high yielding rubber has been carried out in Malaysia with government assistance for both estates and smallholdings. Smallholders were subsidised to the extent of two-thirds of the cost of replanting, and the government made clones and seedlings from its nurseries cheaply available. The cost of the scheme was met partly by a cess on exports. During the 1960s, over half of the total area in rubber, including a quarter of the smallholdings, was planted with high-yielding strains.

Not all former rubber land is, however, being replanted in this way. To reduce the impact of low prices, many estates have diversified their production to some extent, the oil palm being an alternative in Malaysia.

In 1975 the total area planted with rubber was about 2 million hectares with 645,000 hectares under oil palms. Under a new plan the aim was to replant a further 200,000 hectares of rubber and to consolidate many of the smaller holdings. To be economically viable it was considered that a smallholding should have more than 3 hectares of rubber whereas many still had as little as 1 hectare.

The Future of the Plantation System

With the rapid decline in colonial administration and corres-ponding freedom of choice — subject to economic pressures — of the indigenous population, the plantation system had to necessar-ily face major adjustments. Malaysia again provides a good illustration. Since independence, the country has been governed by moderate parties disposed to favour the continuation of the British connection. These governments have guaranteed the integrity of the foreign-based companies and their freedom to repatriate both dividends and capital. The policy of future governments could, however, change. There is inevitably much local feeling against foreign companies, which is not met by present taxation of company profits or even by dues on exports. In addition, in south-east Asia, the persistence of civil wars, armed 'confrontations' and guerrilla warfare discourages foreign investment, particularly in a commodity, such as rubber, which is faced by competition from a synthetic product not liable to such uncertainties.

Political factors weigh almost wholly against the plantation system. Some of the economic factors which also militate against them are of major importance — high charges on capital, management costs and other overheads such as the cost of roads and social services for the workers. On the other hand, the resources of the larger companies permit economies of scale and rapid application of new methods. Economically, therefore, the plantation has features to recommend it. In land use practices, too, there are points for and against the plantation. In the past, many malpractices have been current, partly through companies trying to introduce into tropical environments procedures which were unsuitable, though successful in temperate latitudes. The introduction of clean cultivation between bushes or trees may be cited, this having led to impoverishment of the soil and increased erosion. Whether such errors have led to long-term damage has depended on how serious the damage became, and how quickly remedial measures were adopted. A fundamental disadvantage derives from the fact that plantations generally are devoted to single crops, or at least the cropping programme is orientated mainly to one crop. The dangers of soil exhaustion linked with monoculture are therefore endemic. On the other hand, reputable companies aim to stay in business and to expand, and this they cannot do with impoverished soil and falling yields. Consequently they generally try to keep up soil fertility. This is often partly achieved by growing suitable crops such as nitrogen-fixers as ground cover between the producing plants, which reduces soil erosion at the same time. The larger companies are able to devote resources to research and experiment into the most efficient methods of production. Commercial companies also sponsor independent research organisations, such as the Rubber Research Institute in Malaysia.

Some of the technical and commercial advances made under the plantation system benefit the smallholder, so that gains are diffused through the economy directly as well as through trading needs and monetary expenditure. Nevertheless, it is to be expected that the smallholder will generally be opposed to the plantation economy, which constitutes one of the main political weaknesses of the plantation system in formerly colonial, now independent, countries. The other is that most, if not all, plantation crops can be produced by smallholders, even if not as efficiently as by commercial plantations. Consequently it is hard to avoid the conclusion that the plantation system is likely to decline, and its place be taken increasingly by smallholder production, or by medium-sized estates operated largely without foreign capital. The diffusion of technical aid to developing countries facilitates this process by making the smaller producers more able to take over the role of the foreign companies. Therefore, we may expect to see continued increases, for this as well as other reasons, in the commercialisation of formerly subsistence farmers. That this process has already gone a long way is suggested in Chapter 8, and is a reason for treating subsistence production in this book as a relict form of agriculture, alien to modern forms of production, rather than as a form which must be studied before current systems can be understood.

When a plantation is superseded by another form of organisation, it does not follow that this will comprise smaller units acting independently or linked only through voluntary co-operatives. An alternative form is the collective farm, in which the larger unit may be maintained, though under a kind of peasant ownership, and production directed by an elected management. Most systems of collectivisation have come about in the search for greater efficiency in peasant agriculture rather than in the course of the break-up of a plantation system, but it is an obvious alternative when this occurs. Such a form of organisation may have advantages in making available much of the efficiency of the plantation system without its social and economic disadvantages. Gregor has classed the collective (and state) farm, as developed in the U.S.S.R. for cotton, for example, as a form of plantation, and argues that, in its varied forms, the plantation is likely to remain a major form of agricultural enterprise.[1] Although the continuity between the two forms is clear it is felt by the present author that the differences between the two systems merit separate treatment of state and collective farms.

[1] Gregor (1965).

CHAPTER 7

State and Collective Farming in the U.S.S.R.

In the forms of agriculture so far considered, the units of organisation, whether farms or plantations, are controlled by the individuals who own or manage them. In the plantation owned by a public limited liability company, ownership and management are separated and the function of shareholding does not, of itself, convey the right to participate in the day-to-day management of the company. Even long-term strategy is beyond the influence of most shareholders whose only opportunity of voicing their beliefs is at an annual meeting. Although subject to some form of government influence through manipulation of prices, subsidies, tariffs and market controls, the farm entrepreneurs in public companies are only a little less free than private farmers to make their own judgments on the disposal of their resources, planning of their production, and the sale of their output. The agricultural landscape is therefore made up of a large number of individual units not subject to overall planning control and this is commonly reflected in heterogeneity of production and physical appearance.

In recent decades, however, a very large proportion of the world's agricultural area has come under the direction of state agencies. The U.S.S.R. and China account for most of this state-controlled farming but it applies also to a greater or lesser extent in other countries with communistic forms of government.[1] True state farms — in which ownership and management are vested directly in the state — account for a comparatively small proportion of land and production in most of these countries, the usual form of organisation being collective.

The term 'collective farm' means a holding which is jointly owned or occupied by a number of persons, and operated by them in accordance with a predetermined plan, binding on all the

[1] Jackson (ed.) (1971) provides a valuable review.

144

members of the community. It thus implies a much tighter form of organisation than co-operative farming, in which the individual units remain intact and co-operation is on a voluntary basis, whether this extends to shared working arrangements or only to co-operative purchasing and sales. It differs also from joint ownership or farming of common land in which individual holdings are not fenced, but operation — usually grazing — is on a more or less individual basis.

Collective farms are not confined to communist countries but occur also in Italy, Mexico, India, Pakistan, Japan and other countries. The Israeli *kibbutzim* are probably the most-studied collective settlements, other than those in some of the communist countries. The most fundamental difference between them and the collectives of the U.S.S.R. and China is that the *kibbutzim* are voluntary organisations and are free to determine their own programmes without the necessity of fulfilling government quotas for stated products.

The central idea of the collective farm is the introduction of communal ways of living and working, as a means to satisfying an ideological concept. This concept assumes that greater social satisfaction may be derived from group life than from an individualistic family social structure. The degree of family private life permitted varies from one form of collective to another, as does the amount of property owned individually.[1] In voluntary collectivisation, there is usually an element of asceticism in the belief that the pursuit of private acquisition of property is undesirable. In the collectivisation that was forced on Soviet and Chinese peasants a similar ideology was assumed, as throughout the communist dogma, but it could not be said that there was unqualified acceptance of this ideal by the people who were being organised into collectives. A substantial part of the driving force in collectivisation in the U.S.S.R. and China undoubtedly derives from the previous inefficiency of agriculture, necessitating fundamental reorganisation, together with the advantages in communal forms of organisation for the maintenance of the control by the centralised state and the furthering of government policies generally.

In China, the communist reorganisation of agriculture was pushed ahead much more rapidly than it had been in the Soviet

[1] Digby (1963).

Union. In 1958, the Chinese assumed that full communism could be achieved almost immediately through the institution of the commune. During the turbulent years that followed, the communes underwent many changes but, in general, evolved into large productive and social complexes including not only agricultural but also industrial units. It was reported in 1976 that Chinese rural organisation and agricultural production were divided among 50,000 people's communes and examples with between 17,000 and 82,000 population were described by one visitor.[1] These numbers are much larger than the populations of Soviet collective farms. The areas of land, 2600 and 16,000 hectares in the above cases, were closer to the areas typical of Soviet farms, Chinese lands being generally much more densely populated and intensively cultivated.

Collectivisation in the U.S.S.R.

In the U.S.S.R., in theory, the land of a collective farm (*kolkhoz*) is national state property but is leased permanently by title deeds to the workers of the farm.[2] All the working capital belongs to the kolkhoz except for dwellings, gardens, small personal plots and minor implements. Farm policy and production is decided in accordance with the national and regional plans and applied by the farm committee. Production is subject to state taxation, mainly in the form of stated quantities of produce at fixed prices. As the quantities demanded are absolute and not proportions of production, it is in the interest of the members of the farm to maximise production as they are entitled to sell the balance on the open market. The proceeds are divided among the members according to the nature and hours of work done by the individual. Hired labour is not normally employed.

The collective farm now owns its own machinery, the machine-tractor stations (MTS) which formerly served them having been closed since 1958. The purchase of the machinery by the collectives required considerable outlay of capital and this was in some cases the reason for amalgamation of farms. The possession of their own machinery by the farms has eliminated the difficulties that had arisen from control of equipment in other hands, including the meeting of dates on which the machines

[1] Rey (1976).
[2] Kolkhoz is an abbreviation of *kollektivnoye khozyaystvo* (collective farm).

were required by individual farms. The training of farm personnel over the years had eliminated the need for the MTS for routine maintenance, but major repairs are carried out by stations operated by the Farm Technical Service (*Sel'khoztekhnika*).

The elimination of the MTS also meant the end of payments of substantial portions of produce which had to be made by the farms to the MTS for their services. At the same time further reforms raised the prices paid by the state for the compulsory deliveries of produce, which had formerly been artificially low to keep down the cost of living in the towns.

The members of the collective farms retain their personal plots which have been permitted since the early days of collectivisation. At times efforts have been made to reduce them, and it has been said that the long-term aim is so to improve the output of the collective farms and the income of the farmer that the personal plot will become unwanted. However, it continues to have lavished on it the maximum amount of attention and fertiliser that the farmer and his family can manage. Although limited to a maximum of one-half of a hectare, the personal plot has carried the family through very difficult times with intensive production of potatoes, vegetables and fruit for home consumption, and maize and roots for the personally-owned cattle, pigs and poultry.

In the three-year period 1972–74, one-third of the meat and milk and 62 per cent of the potatoes and 44 per cent of the eggs produced in the U.S.S.R. came from the personal plots which accounted for only about 3 per cent of the land in cultivation.[1] Previously the proportions were much higher but the decrease has continued steadily as improvement has been effected in the public sector of the agricultural economy. Yields from the personal plots continue to be much higher than from the same unit area of the collective land, but it must be remembered that the crops sown in them are those which respond to intensive cultivation, and the vegetable garden of a commercial farm will also normally yield more highly than the broad acres in mechanised cultivation. The produce of the personal plots is consumed mainly by the users but surpluses are sold in kolkhoz markets located in the towns. Shoppers tend to prefer them to the state shops because the produce sold in the markets is fresher.

[1] All statistics are from annual volumes of *Narodnoye khozyaystvo SSSR*.

There has been an overall trend towards larger units since the beginning of collectivisation in 1928. The 25 million peasant holdings of 1928 which had averaged about 4 hectares of arable land had become by 1940, 235,500 collectives. By 1950 amalgamations had halved the number of units to 123,700 and further amalgamations, together with conversion to state farms, reduced the number of collectives to about one-quarter of this number by 1974, viz. 30,700. The collectives then averaged 3400 hectares of arable land and 463 households. The enlarged collectives of the present day offer more scope for efficient cultivation of more diversified crops with maximum mechanisation and electrification. In 1958, only about half of the collective farms were electrified, but by 1973 the number using electricity had risen to 99 per cent.

In spite of the reduction in the area cultivated collectively from 105 million hectares in 1965 to 98·6 million in 1973 numbers of livestock had been increased. A smaller proportion of each class of livestock is, however, found on the collectives as the state farms have increased their share. Numbers of livestock on personal plots do not show any marked changes except for a sharp decrease in goats and an increase in pigs, a move which suggests better husbandry.

The traditional settlement pattern is one of nucleated form, with the individual houses grouped near but separated from the communal buildings of the kolkhoz — the farm buildings as well as the general store, school and community hall. These vary in pattern, design and materials of construction as well as in location on valley bottom, interfluve or mountain slope, according to region.[1] The new buildings, as commonly in other parts of the world, show less regional variation, but respond to the economies of building in local materials and siting in accordance with natural advantages. Some of the enlarged collectives have been given new centralised villages in order to facilitate provision of better services.

State Farms

The state farms are owned and operated by government agencies and have been generally larger and better endowed with equipment and capital than the collectives, and it has been easier

[1] A very useful summary is given by Pierre George (1962), 297–301.

to make them efficient. State farms include experimental and special farms but most are essentially devoted to normal production, the name *sovkhoz* being applied to this type of farm.

The average sovkhoz is probably hardly a more valuable concept than the average farm in other countries, but a few figures for 1974 do show the great size of these farms: thus, average labour force, 570; average cropped area, 6000 hectares. The average grain sovkhoz had more than three times this cropland — 19,900 hectares. On this type of sovkhoz the livestock averaged, cattle 3674; pigs 1949; sheep and goats 4471. To quote only one more comparison, the sheep sovkhoz averaged over 30,000 head of sheep and goats on a total area of about 114,000 hectares. Even these figures show a slight reduction of average size from the peak figures of a few years earlier, resulting from the reorganisation of more somewhat smaller collectives as state farms.

In spite of amalgamation of collectives to form larger units, presumably in search of greater efficiency, many have been made into state farms. The regional distribution of lands as between kolkhoz and sovkhoz as it was some years ago is shown in Figure 8. Since the data for this map were compiled, the state farms have grown still more and in 1974 accounted for an area of agricultural land 84 per cent greater than that of the collectives. In terms of cultivated area, in 1974 the state farms accounted for 10 per cent more than the collectives.

A rise in importance of the state farms in grain production from 11 to 41 per cent between 1950 and 1963 reflected the extension of cultivation in the 'virgin lands' of west Siberia and north Kazakhstan and for pioneering cultivation in other areas. In the five years 1970–74 the average contribution from the state farms was 46 per cent of the total grain harvested.

The *sovkhozi* are the principal means of extending the frontiers of cultivation and of promoting advanced methods of agriculture. In the last few years they have been increasingly used to develop livestock husbandry. In the Baltic provinces, for example, the emphasis has been on dairying, and around Moscow, on dairying and horticulture. Altogether in 1974, 6814 state farms were devoted principally to dairying and beef production, 1385 mainly to sheep, 952 to pigs compared with 1579 to grain, 2643 to fruit and vegetables and the remainder to sugar beet, cotton, etc.

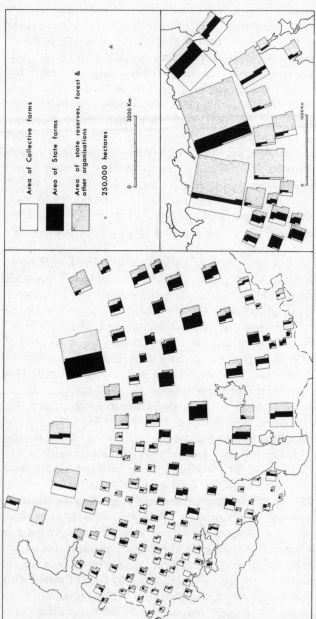

Figure 8. State and collective farm lands in the U.S.S.R. Note the higher proportion of state farms in the areas of recent development in Kazakhstan, western Siberia, central Asia and some northern regions, but in some of these regions of pioneering farming, forests remain dominant and suggest the difficulties of further agricultural expansion.

Source: *Atlas razviitya khozyaystva i kul'tury SSSR*, 1967.

Altogether they numbered 17,700 and covered 355 million hectares (nearly 900 million acres) of agricultural land.

In terms of livestock products marketed through state outlets, the contribution of the state farms rose from 21 per cent in 1950 to 44 per cent in 1973. In milk and wool the increase in the period was from 15 per cent to over 40 per cent, and in eggs it was from 7 per cent to 71 per cent.

In areas of sparse settlement, such as the mountain regions of the Far East, where collectivisation is impracticable, a small number of individual farms remain as virtually private holdings, but they are now insignificant in the total scene. Their continued attenuation emphasises the thoroughness of the transition in farm organisation in the Soviet Union.

Historical Perspective

In order to assess the achievements of Soviet reorganisation of agriculture, it is necessary to consider the antecedents of the present system, the most important elements being the late survival of serfdom, continued exploitation of the peasantry, land hunger and the absence of a revolution in agrarian methods comparable with that which took place in western Europe before the twentieth century.

In medieval Russia there were many categories of more or less enslaved peasants. In the sixteenth and seventeenth centuries, when in England the conditions which produced the agrarian and industrial revolutions were being created, in Russia the landowners were consolidating their control over the peasantry. Ancient customs, new economic pressures and increasing control of the land by the tsars through the aristocracy facilitated the binding of the peasants to their masters. Peasant migration was a feature of the sixteenth century, which was marked by internal struggles, destruction of villages and laying waste of land, especially in the central parts of Muscovy. The rights of free peasants to leave landowners provided that their dues had been met was increasingly challenged and the landowners demanded increased rights to pursue runaways. New land grants to the middle classes and military gentry who had helped found the Romanov dynasty in 1613 were followed by the code of Tsar Alexis in 1649. This abolished time limits which had previously limited the recovery of runaways.

Associated with the practice of serfdom was the commune, or *mir*. The origins of the commune are lost in pre-history, but it seems clear that widespread collective arrangements had evolved in peasant communities for regulating allocation of land, land utilisation, admission of newcomers to the group and other local affairs. The commune was used by the central government to assist in the calculation and collection of taxes.

B. H. Sumner gives the following description of the commune in the nineteenth century:

> At the time of the emancipation most of the Russian peasantry were grouped in communes, of varying sizes, composed sometimes of one village, sometimes of parts of one village, sometimes of groups of scattered settlements. The most essential usual features of the commune from the agricultural and economic standpoint were: (i) that its membership was hereditary, though newcomers could be admitted; (ii) that its members worked land by families, but (except in a minority of communes where holdings were hereditary) periodically redistributed their strip holdings scattered in the (usually three-course) open fields, in accordance with either working strength, or taxation and other obligations, or the number of 'eaters' in each family; (iii) that the members of the commune regulated in common the use of meadows, pasture, fisheries, woods, etc., and the disposal of any communal land not already utilised and the acquisition of new land or working rights.[1]

When at last the emancipation of the serfs came in 1861, in response to economic and social pressures from industrialists and the more progressive landowners who wanted a more efficient labour force, as well as from the largely inarticulate body of the peasants themselves, the reorganisation was based on the *mir*. Although the freed serfs had to pay for their holdings to the state, mainly by redemption annuities for 49 years, to repay the compensation given to the landowners, the title to the land was vested not in the individual but in the *mir*. This, together with the high rates of the annuities, the large areas the landowners were allowed to retain and the grazing and forest rights that were given to the landowners, resulted in the peasants feeling that they had

---
[1] Sumner (1961), 136.

again been cheated. Emancipation was therefore followed by further unrest and rioting, which was quickly suppressed in the usual brutal manner developed by the tsars and their agents.

The commune retained its old functions and added new ones, including collection of the redemption annuities. It stood between the peasant and the state and, although it represented a measure of peasant self-government, it was the object of much of the peasant's dissatisfaction because it administered so much that was unpopular. Meanwhile, the lands worked by the peasants remained for the most part in the three-field system and the strip holdings were generally still redistributed at intervals. In the areas where population pressure was highest, in the central black-earth and middle Volga provinces in particular, the peasants found themselves with far too little land. Here and in other regions, the more fortunate — and the more commercially-minded — were able to buy out their less fortunate neighbours and enlarge their holdings. Thus arose the *kulak* class, later the object of the hatred of the communists. The population grew more rapidly and the pressure on the land became more intense. The emancipation had in fact done little to improve the lot of the peasants. They even still had to have a passport (as they had since the time of Peter the Great, and now granted at the discretion of the commune) before they could move about. The commune, because of the manner of levying the poll-tax and other taxes, had reason to try to prevent men leaving it.

Except in Siberia, in which selfdom had not been fully developed, and the newly colonised lands in the European and Asiatic south, the remnants of serfdom were still clear twenty years after the emancipation edict. Continued unrest culminated in 1905 in widespread riots and arson. Thus, peasant unrest was contemporaneous with, though not integrated with, the revolts in St Petersburg and Moscow which were the main expression of the Revolution of 1905. This secured the 'October Manifesto' including the establishment of the Duma, a parliament which, though it had little power, signified the weakening of tsarist autocracy. But the peasant riots continued in 1906. Severe repression by the military and the execution of some 4000 persons was followed, belatedly as always in tsarist Russia, by some measures of agrarian reform.

The 1906 reforms were designed to enable the scattered strip holdings to be consolidated into more efficient farm units. The hitherto ineffective Peasants' Land Bank, set up in 1883, was improved to enable peasants to rent additional non-peasant land. The control of the peasant by the commune was reduced by the abolition of communal responsibility for taxation. It was made easier for peasants to leave the commune and to migrate to the new lands in Siberia and the south.

During the next ten years considerable progress was made in the modernisation of agriculture. More machinery and fertilisers helped the expansion of the cropped acreage. After the crop failures and famines of 1905–7 there were a number of good crop years. Even so, in 1917, the majority of the peasants were still living in a deplorable state of poverty, many of them bound to landlords, no longer as serfs but by debt. The situation was still particularly acute in the central black-earth and middle Volga provinces, where population pressure was heaviest. The majority of peasants in European Russia were still tied to the communes, their land in scattered strips in the open fields. Only about a tenth of the peasant households had been actually resettled on the new consolidated holdings. The success of the more prosperous peasants, the kulaks, in buying up holdings vacated by peasants leaving the communes, and in renting land from the gentry, did nothing to assuage the land hunger and dissatisfaction of the majority. The war, with its inevitable drain of manpower from the countryside as well as the towns, and the appalling casualties — 40 per cent of the Allied death roll was attributed to Russia — did not divert the peasants from the age-old conviction that the gentry had stolen their land, and the traditional faith in the tsar as the little father who would protect them from their immediate superiors had worn a little thin.

Agrarian Reform in the Revolution

In the deliberations of the various organisations that worked towards revolution in the latter part of the nineteenth century and until its achievement in 1917, the question of agrarian reform was, of course, well to the fore. It was intimately bound up with the assessment of the part that could be expected of the peasants in a large scale uprising. Plekhanov, the first Russian exponent of Marxism, discarded the romantic view of the *mir* or commune as evidence of the peasants' adherence to a communal way of life, stressing the need to focus revolutionary organisation on the industrial proletariat. This class was growing rapidly and was being exploited mercilessly in the factories and mines. Lenin, however, while accepting the role of the industrial workers as fundamental, did not reject the peasantry as uniformly conservative and counter-revolutionary. His first period of exile, for participating in student demonstrations, was to a rural district and his first writings were on the situation of the peasants.

After the split of the Russian Social Democratic movement into Bolshevik (majority) and Menshevik (minority) wings in 1903, Lenin pursued the thesis of joint industrial and rural proletariat action. Their combined congress of 1906 adopted the Menshevik programme of agrarian reform, with land to be held by the *zemstvo*, or district council, and leased to the peasants, but the Bolsheviks favoured collective cultivation with state aid. As already noted, the fortunes of the peasantry had improved only marginally by 1917 and when, eventually, military defeat and food shortages led to revolution in St Petersburg in March, the provisional government that took office had still to face the need for urgent agrarian reform, as well as labour legislation for the industrial workers, and the continued prosecution of the war.

Side by side with the provisional government formed by the Duma, and competing with its authority, soviets or councils of workers and soldiers were formed and similar groups were later formed in the rural areas. Thus, there were already in being bodies through which orders could be passed to the countryside when the Bolshevik *coup* of October 25 (November 7 new calendar) gave them control. One of the first two decrees provided for termination of private ownership of land and its immediate distribution to the peasants and workers, the other concerning the ending of the war. In fact, the peasants had already begun to seize land and were not to be easily won over to the Bolshevik cause. Their retention of food supplies worsened the supply situation in the towns and drew upon them the force of the Red Army and the secret police, re-created by the Bolsheviks. Thus the first days of the Soviet government did not augur well for relations between the new authorities and the peasants. But worse was to follow with the development of the 'white' counter-revolutionary movement and the intervention of the Allies in support. For three years armed struggle raged from the

Baltic to the Bering Straits. The French organised the particularly bitter fighting in the Ukraine, Czech troops held the Trans-Siberian Railway, thousands of Japanese troops were poured into the Far East and British, American, Polish, Greek, Bulgarian and other regular and irregular forces took part. The peasantry gave up the hopeless struggle to cultivate crops which were likely to be forcibly requisitioned if they were not destroyed in the fields. Near-famine conditions prevailed in many rural areas as well as in the towns. By the winter of 1921–22 reserves had been used up and the famine of that time was estimated to have caused the death of some five million people, and many more would have died but for food distributed under international aid arrangements.

Shortly before this deterioration, Lenin in his New Economic Policy had accepted the need for compromise with capitalist interests which could revive the economy. In the following years, when the peasants were allowed considerable freedom in their sales of produce, they took the opportunity of getting high prices and the kulak class recovered and prospered. Meanwhile Lenin died and the struggle for power within the Communist Party intensified, resulting in the destruction of all opposition by Stalin. In December 1927 the Fifteenth Party Congress sanctioned his programme for forced industrialisation and agrarian revolution, based on collectivisation.

Progress in Collectivisation

The Five-Year Plan, accepted in 1928, required collectivisation of up to one-fifth of the farms by 1933. It was stated that the kulaks had withheld grain deliveries, thus re-creating conditions of food shortage, and the raids on kulak farms, requisitions and arrests which began in 1928 were intensified. In the next two years this class, estimated at $1\frac{1}{2}$ to 3 million men, was ruthlessly destroyed, and many of the much larger group of middle-class peasants were also arrested and deported to labour camps. In their resistance the peasants hid or burnt crops, slaughtered livestock and wrecked machinery. Stalin himself saw the folly of this brutal attack and ordered an end to force, but not until 1930, by which time the damage was done to the cause of collectivisation and the economy further weakened. Later it was revealed that between 1929 and 1933 livestock numbers fell by some 30 million cattle, nearly 100 million sheep and goats and 17 or 18 million horses. Stock raising had always been less well managed in Russia than crop cultivation and it was now further weakened. Even nomadic and semi-nomadic Kazakhs reduced their stock.

In spite of the slowing down of the programme, by the end of the first Five-Year Plan in 1933 about 60 per cent of all holdings had been collectivised as compared with the 20 per cent planned before the intensification of the drive.

The original collectives were formed by the amalgamation of about 75 peasant holdings administered by a chairman and his assistants. The workers were required to devote 100 to 150 days to the co-operative enterprise. The state at that time claimed all produce over the allowances for the subsistence of the collective.

An essential prerequisite for the successful cultivation of the new large farms was mechanisation. The urgency was increased by the losses in horses. Only 7000 tractors were in use at the beginning of the drive but in 1929 alone 30,000 were manufactured. The machine-tractor stations (MTS) were introduced to act as pools for machines, operators and mechanics and by 1936 there were almost 5000 of them. This relieved the kolkhoz management, already overburdened with the task of training the work force, of the problems connected directly with mechanisation, and provided powerful units of workers on whom the Party authorities could rely throughout the countryside.

Between 1931 and 1933 famine struck again, partly because of the resistance of the peasants to the collectivisation. Among the worst-hit regions were the most fertile — the black-earth lands of the middle Volga, Ukraine and the northern Caucasus regions. It is said that between ten and eleven millions fell victim. In 1933 Stalin relaxed his policy further and allowed grain to be withdrawn from stores and purchased abroad. Concessions were made to the kolkhoz labourer, most important being the allocation of a small plot of land which he could cultivate for his own purposes. He was also now allowed to keep his own cow, a few sheep and goats and some poultry, and receive a cash share in the proceeds of the sales of the kolkhoz.

The kolkhoz was now favoured at the expense of the sovkhoz and many of these great farms which had also been created during the preceding years were broken up and the land distributed to collective farms. Others were retained as research, training and stud farms. As already noted, this trend has since been again reversed, the sovkhoz being now the major institution in the development of Soviet agriculture.

By 1940, as already noted, the 25 million holdings of 1928 which had averaged about 15 hectares had become 235,500 collective farms, averaging about 1600 hectares in total size. On these lived about 75 million people with perhaps another million serving on the machine-tractor stations. By this time the collective had been able to demonstrate some advantages in working the land, especially in the more difficult areas. In arid and semi-arid conditions sowing could be completed quickly, when conditions were most satisfactory, irrigation facilities could be used more efficiently and new cultivation could be planned to best advantage. In addition, economy of labour released large numbers of workers needed in the towns and on construction projects.

During the war, which raged on Russian territory from 1941 to 1945 and caused loss of life amounting to some 20 million people and the destruction of 25 million homes, Soviet agriculture suffered a further immense setback. The 'scorched earth' policy which denied the advancing Nazi troops food, shelter and machinery meant incredible sacrifices for the long-suffering peasants. After the war a renewed drive had to be made to modernise production. In 1951 it was claimed that almost all ploughs were mechanically operated and 60 per cent of the harvesting was by machine.

Between 1949 and 1951 an experiment was initiated with enlarged collectives, formed by the grouping of three to five existing collectives. By 1951 the 235,000 collective farms had been combined into half this number. Consolidation of collectives into fewer farms has continued together with conversion into state farms as recorded above.

Regional Variation

The vastness of the Soviet Union results in wide divergence of physical conditions. Cropping is impossible over most of the north because of cold and much of the south because of aridity. The great mountain ranges also prevent cultivation or restrict it to favoured valleys. Where cropping cannot be practised extensive pastoralism may be possible, but the generally dry conditions prevailing throughout most of the Soviet Union result in the

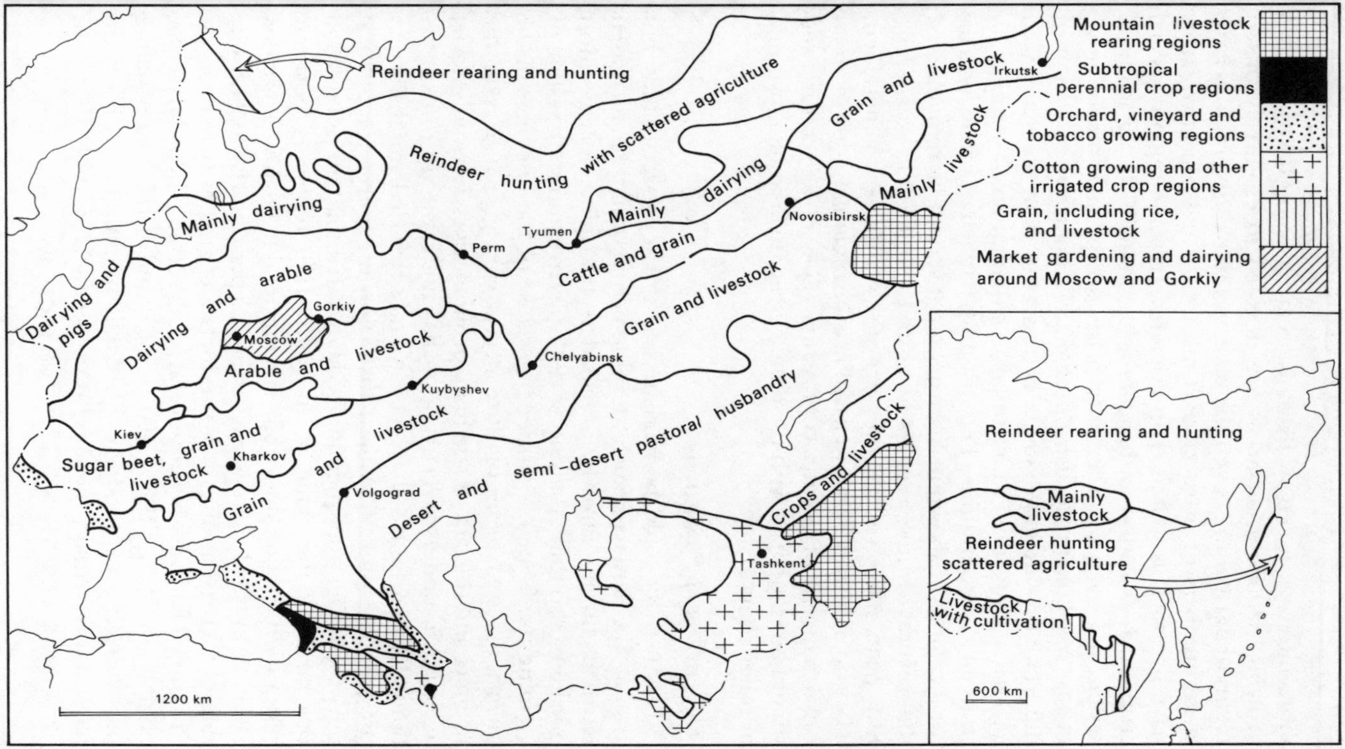
Mountain livestock rearing regions
Subtropical perennial crop regions
Orchard, vineyard and tobacco growing regions
Cotton growing and other irrigated crop regions
Grain, including rice, and livestock
Market gardening and dairying around Moscow and Gorkiy
Reindeer rearing and hunting
Reindeer hunting with scattered agriculture
Grain and livestock
Irkutsk
Mainly livestock
Mainly dairying
Novosibirsk
Tyumen
Perm
Cattle and grain
Dairying and pigs
Dairying and arable
Gorkiy
Moscow
Grain and livestock
Arable and livestock
Chelyabinsk
Kuybyshev
Desert and semi-desert pastoral husbandry
Crops and livestock
Kiev
Sugar beet, grain and livestock
Kharkov
Grain and livestock
Volgograd
Tashkent
1200 km
Reindeer rearing and hunting
Mainly livestock
Reindeer hunting scattered agriculture
Livestock with cultivation
600 km

Figure 9. Agricultural regions of the U.S.S.R. (after A. N. Rakitnikov and others).

more intensive forms of livestock rearing being based on crop growing. Hence great areas are devoted to essentially mixed forms of agriculture. Even the main grain-growing regions have large numbers of livestock and most state and collective farms have livestock as well as crop brigades. In some regions special crops such as cotton and sugar beet dominate the agricultural system but even so other enterprises are present on the huge farms. Market forces also operate to influence the location of different production systems because the state seeks to avoid excessive costs due to unsuitable exploitation — subject to strategic considerations. The resulting pattern of broad agricultural regions is shown in Figure 9 but it should be remembered that several of these regions cover areas larger than most entire European states so that they are necessarily very generalised. They do, however, serve well to illustrate regionalisation in broad zonal patterns of greatly contrasting types of agriculture and should be compared with the detailed regionalisation illustrated in other chapters.

The Problem of Productivity

Productivity could be increased by greater specialisation but among the costs of such developments would be increased burdens on an already overloaded transport system. Most outsiders also think that the Soviet collective and state system of farming itself hinders efficiency because of the hugeness of the farms, the vast army of bureaucrats who give orders to the farmers and the lack of adequate profit motives. Certainly Soviet agriculture under collectivisation has not come up to expectations. Official figures have consistently revealed unsatisfactory progress in spite of evident distortion of figures for political purposes. The published index[1] shows that total output in 1950 was virtually the same as in 1940 (100) following post-war restoration. It took until 1969 for this total to be doubled (index 201) but by 1973 there was a further 25 per cent rise to 249. The rate of increase in livestock products was greater than in crops. Prolonged bad weather in several years resulted in bad harvests, notably in 1957, 1963, 1965, 1967 and 1972, necessitating purchases from abroad. At 168·2 million tonnes the 1972 grain harvest was virtually the same as the 1966–70 average (167·6). In

[1] *Narodnoye khozyaystvo SSSR v 1974 godu* (1975), 307.

1973 a new record of $222 \cdot 5$ million tonnes was claimed but the following year the harvest dropped to 196 million tonnes. The drought year of 1975 saw a further dramatic fall back to 160 million tonnes. In 1976 there was a recovery to over 200 million tonnes but the failure to achieve reasonable stability is evident.

It is, of course, generally recognised that physical conditions in the Soviet Union are far inferior for agriculture generally to most of the U.S.A. or western Europe. Even in the 10 per cent or so of the U.S.S.R. which is within the limits set to cultivation by cold, drought and relief, the more humid areas have long winters and the warmer areas are mostly dry.[1] In such circumstances yields can never compare with those in more favourable conditions in western Europe and the U.S.A. under similar systems of cultivation. Where Soviet agriculture seems to have failed relatively is in the failure of the most fertile regions to produce as highly as might be expected, particularly with comparatively high inputs of labour.

Compared with the previous decade, however, the position shows substantial improvements. Table 5 compares the 4-year averages for 1971–74 with the 5-year averages for 1961–65 and 1966–70. It is at once apparent that major advances have been achieved in livestock husbandry, long regarded as the greatest weakness in Russian agriculture. This is supported by individual advances, such as in yield of milk per cow, which for the whole of the U.S.S.R. is recorded as averaging 2242 kg in 1973 compared with 1137 in 1950 and 1853 in 1965. The increase has been especially notable in the collective farms, for which the average yield since 1973 has been virtually the same as recorded by the state farms, which had always previously returned higher figures. In fact, about one-half of the increase in the amount of milk produced since 1965 has resulted from increased yields, the other half from an increase in the national herd. Yields per cow are still far below those achieved in the more advanced dairying countries such as New Zealand and parts of western Europe. In the case of grain production, whereas the increases in the 1950s came mainly from the ploughing-up of 40 million hectares of virgin lands in Kazakhstan and western Siberia, there was no increase between 1965 and 1974 in the area sown to grain. The wheat and rye areas actually contracted but there was a corresponding increase in the

[1] Symons (1972), Chapter 4 examines the problems in detail.

area under barley and oats, reflecting both the demand for animal fodder for the enlarged herds and a more realistic appraisal of the best use for the croplands of the cooler and drier parts of the Soviet Union.

TABLE 5

U.S.S.R.: PRODUCTION AVERAGES 1961–74

	million tonnes		
	1961–65	1966–70	1971–74
Grain	130·3	167·6	191·9
Cotton	5·0	6·1	7·6
Sugar beet	59·2	81·1	78·4
Potatoes	81·6	94·8	90·0
Vegetables	16·9	19·5	22·9
Meat	9·3	11·6	13·8
Milk, etc.	64·7	80·6	86·6
	000 million		
Eggs	28·7	35·8	49·9

Source: *Narodnoye khozyaystvo SSSR v 1974 godu*, Moscow, 1975.

As a result of the changes in cropping policies, the increasing availability of fertilisers and improved strains of crops and cropping techniques, yields of all the main crops have improved.

Taking the Soviet population as 250 million (1973 estimate), the average grain production per head for 1971–74 was 770 kg, compared with 570 kg for the smaller population of 1961–65 and 360 kg for 1946–50. More of the recent production, however, is used as animal feed.

For the improvement of production one of the fundamental requirements has been continuing increase in the fertiliser available. During the 1950s mineral fertilisers were in very short supply but the capital investment programme initiated in 1963 resulted in an increase in total production from under 14 million tonnes in 1960 to 55 million tonnes in 1970 and 80 million tonnes in 1974. This made possible an application of 287 kg per hectare of cultivated land, more than twice the 1965 figure, but there were still many complaints of poor quality, inadequate packing and damage in transit of fertilisers. The means of application is interesting. In 1965 one-third of all mineral fertiliser applied was distributed by hand, one-third by ground machine and one-third from the air. The Soviet agricultural aviation industry is the largest in the world and nearly one-half of the country's cultivated land is treated each year from the air, either with fertilisers or protective crop sprays.

Mechanisation in general is, however, still lagging behind needs. Although about 350,000 tractors are delivered annually to Soviet farms and the total stock of tractors exceeds two million, there is an overall shortage and many are out of operation at any given time owing to shortage of spare parts or repair delays. Although the agricultural technical service is believed to be a vast improvement on the machine-tractor stations there are many complaints about the quality of the service. A similar situation applies in transport services. Farms are required to use the railways wherever possible but these are overloaded and complaints about delays and shortages of wagons are common. Many cases of losses subsequent to harvesting for these reasons are reported every year. On the other hand, it must be recognised that the annual task of movement of the grain harvest alone, over the vast distances of the Soviet Union, represents an enormous task. It is one of the drawbacks of specialisation of area that the transport of produce to the more densely populated areas is increased. The development of the virgin lands in the eastern regions has placed great burdens on the Soviet railways, but it has spread the risks in the harvest and released Ukrainian and other western areas for more balanced agriculture, especially for livestock products which are in great demand in the cities.

Another road to increased production is by way of land improvement. Even though much drainage work has been carried out it has been estimated that still more than two million square kilometres of the Soviet Union could be improved in this way. Waterlogging, following the melting of the snow cover, in particular, delays cultivation in the spring. After the virgin lands programme there was a marked shift of resources to intensification of production, especially in western areas, and drainage of the Pripyat marshes of Belorussia and adjacent parts of Russia, the Ukraine and the Baltic republics was stepped up. In most of the southern parts of the U.S.S.R. and especially in Central Asia, however, the need is for irrigation. Nearly 5 per cent of the cultivated area of the U.S.S.R. is irrigated, and about half a million hectares are added annually to the irrigated area. Cotton is the main crop of the irrigated lands in Central Asia but in the Soviet Union as a whole grains and roots are the main irrigated crops.

It is not sufficient merely to extend the area under the plough, under irrigation or other means of intensified production; conservation is also necessary. This lesson was learnt — or re-learnt — the hard way when the newly-cultivated virgin lands lost much of their topsoil and urgent conservation measures had to be taken. Some degree of stability appears to have been attained since the late 'sixties, with more varied crop rotation and support of livestock alternating with grain production in these lands which are acutely prone to drought and the scorching *sukhovey* winds. Irrigation introduces problems of secondary salinification and the drop in the levels of water bodies like the Caspian and Aral Seas because of the tapping of the rivers flowing into them. The Soviet scheme to divert north-flowing Siberian rivers to supplement water supplies in Central Asian territories has led to international controversy. Less spectacular measures to conserve moisture include widespread planting of shelterbelts to reduce wind velocity and evaporation and to prevent snow blowing off the fields where it is needed on to roads and railways.

Apart from physical and technical methods, the social and economic climate on the Soviet farms has had to be improved in the battle to increase production. The Soviet farm has been traditionally organised in brigades responsible for various tasks such as field cultivation, mechanised operations, orchard care, etc. As the farms have grown larger, so the size of the brigades has grown to resemble whole collective farms of a few years earlier. This has given rise to organisational problems and the introduction of various measures to restore individual responsibility and interest, such as the creation of small groups called *zveny* or links, attached to particular areas for the whole period of the growing of the crop. Some see this as a useful compromise within the socialist system, others as a threat to the concept of the collective farm because the attachment of a particular person to a given plot of land has associations with the peasant type of farming which collectivisation destroyed. In any case, economic accountability of brigades and groups, as well as of the complete farms themselves, has been greatly increased in recent years, and the profit motive has become largely accepted as a means of inducing greater efficiency and productivity. Farms in less productive areas are, however, to some extent helped by regional variation in the prices for state procurements.

Payment to farms for their produce has been greatly improved since the early 'sixties, and as a result the remuneration of members of collective farms has been steadily increased, closing the gap between them and the state farm workers whose wages are fixed by the state. The cash element in payments has been increased and minimum earnings and pensions are guaranteed by the state. In spite of these reforms the Soviet Union continues to experience severe labour difficulties in agriculture, with excessive migration from some regions to the cities, but over-population in others.

In conclusion, the Soviet approach to solving the geographical problems of agriculture in a severe environment with additional handicaps inherited from the social organisation of the past has been bold and dramatic but it cannot be said that it has been particularly successful in terms of productivity. The Soviet Union is, however, completely committed to this form of agricultural organisation and it seems likely that the trend will be to yet larger units, with state farms continuing to replace collectives. Changes are occurring along the lines of making the farms more industrial in organisation and operating character with an increase of inter-farm co-operatives for some purposes but the basic struc-ture of Soviet agriculture now seems well established with a likelihood of relatively gradual evolution of the organisation of Soviet farming compared with the revolution of the past.

CHAPTER 8

Subsistence Agriculture

Although production for personal subsistence has at times temporarily increased in some areas because of war or other interruptions to supply, the secular trend has been, and almost certainly will continue to be a decline in the importance of subsistence farming.

Commercial and state-directed systems are so dominant that agriculture is producing primarily for sale in all but the most backward and remote areas. It might be argued that true subsistence agriculture could be said to occur only where there is no exterior trade. Just as, however, a farmer producing essentially for the commercial market will use a little of his output in his home, any practical interpretation of subsistence agriculture must allow of at least some trade. At the simplest level, this may merely be trade by barter between nearby communities occupying areas of somewhat different physical conditions, and so having an incentive to exchange surplus goods. In these later decades of the twentieth century, however, it may be said that commerce has affected most societies and few do not produce some article for sale outside the immediate locality, using the proceeds to purchase at least a small range of manufactured goods and perhaps some foodstuffs to supplement the local produce. How then shall we decide when to refer to an agricultural economy as 'subsistence'?

Whittlesey regarded intention as critical:

> No farming region lives wholly without exchange of surplus, but the percentage of goods exchanged in the total produced ranges from almost nothing to very nearly everything. The critical difference is the intention of the farmer. If he grows his crops or raises his animals with the object of selling, he

166

is a commercial farmer; if he merely sells what he happens to have left over, or what he is forced to part with by emergencies, he is a subsistence farmer.[1]

This attempt at a definition is unsatisfactory because large numbers of producers cater deliberately for both the subsistence needs of their own families and for a commercial market, i.e. the division of intention is not as simple as Whittlesey visualised. Furthermore, the divisions have almost certainly become more blurred since Whittlesey wrote, over 40 years ago, during which time the social, political and economic emergence of the peoples of Asia and Africa has been a world issue. People engaged in all types of farming have been affected by the developments of transport and other facilities for trade which have occurred throughout the world. An example is the Chimbu people of the central ranges of New Guinea, whose society and agriculture have been effectively studied by a geographer and an anthropologist working together.[2] The Chimbu became known to the outside world only in 1933 and for centuries their only link with lowland New Guinea had been tenuous routes by which food crops, traditional feather and shell valuables and, latterly, worn steel knives and axes were traded. Since the arrival of the Europeans, and on a much larger scale since about 1950, money has become important. Small sales of foodstuffs to Europeans provided the main income until about 1959 but sales of coffee then increased rapidly. Unfortunately, while this cash crop and, since 1972, the incursions of tourists, have weakened the old subsistence economy, the addiction of the Chimbu to traditional large-scale periodic celebrations, based on the slaughtering of immense herds of pigs, has conflicted with coffee growing and productive investment has fallen away.[3]

More thoroughly revolutionary in their impact upon traditional methods and objectives of agriculture, as well as politically, are the great changes resulting from the extension of communism. When Whittlesey wrote, rural Russia was in ferment with the aftermath of the first collectivisation programme and to the rest of the world the situation was far from clear. Now, virtually all of the peasants of the U.S.S.R. have been brought into the state

[1] Whittlesey (1936), 211.   [3] Brookfield (1973), 138.
[2] Brookfield and Brown (1963).

economic system. Still more significant, however, is the social revolution in China. It may well be that the majority of Chinese farmers are, in effect, still producing largely for consumption by their own families, but only as part of the efforts of the production brigade and commune. The state has a claim on this produce and aims to derive larger surpluses from every unit to improve the food supplies of the non-agricultural areas and to stimulate the whole economy.

These great changes have reduced to a fraction the former large percentage of the world's population whose farming was based on the intention to provide for their own subsistence and sell only whatever they might 'happen' to have surplus.

In seeking a definition of subsistence agriculture, it seems more profitable to have regard to actual disposal of crops rather than to intention, and to consider cash sales relative to subsistence consumption, taking one year with another.

A working definition, then, might be: 'a subsistence farmer is one whose output is consumed almost entirely in his own home, not more than a small proportion being regularly offered for sale.'

At once another question must be posed — what is the critical proportion that may be offered for sale without the status of the farmer being considered to change? Further, it may be objected that whatever arbitrary figure is selected, only rarely will statistics be available to enable the classification to be used with precision. It is, of course, with the regions and economies about which least statistical information is available that we are here concerned.

One possibility might be the restriction of the term 'subsistence farmer' to one whose contribution to commerce is so small as to be barely measurable. To put this in terms of consumption, such a family unit will have very few purchased goods, food or services, but perhaps odd tools, cooking utensils and matches. This rather narrow definition would exclude large numbers of sedentary cultivators, especially those cultivating, as a minor interest, a cash crop such as cotton or rubber for sale to a marketing agency, and the intensive rice growers whose small individual surplus collectively helps to feed the cities and plantation workers of south-east Asia. Pending any more satisfactory term, the latter folk might be referred to as 'quasi-subsistence farmers'. When cash crops become of substantial importance in a holding which is not yet devoted preponderantly to them, the term 'semi-subsistence farming' would seem to be appropriate.

These classes may be compared with those suggested[1] by an administrator concerned with levying taxation in East Africa, namely (a) pure subsistence — no cash crops; no tax; no import or export of labour, (b) subsistence with taxes — some cash crops grown or employment sought elsewhere mainly for the purpose of paying taxes, e.g. the peoples of Ruanda Urundi, the Karamoju and the Masai. These two groups might both be considered as subdivisions of 'subsistence farmers.' (c) Subsistence plus cash crops, where taxes have become a minor element of cash outlay, which might be equated with my quasi-subsistence group, (d) subsistence plus cash, suggested with particular regard to men travelling outside their district to seek work, (e) agricultural labour working mainly for wages on plantations and (f) wage labour in an industrial economy.

The last two classes are not relevant to the immediate question but the 'subsistence plus cash' group may be equated with the mainly sedentary farmers envisaged in the term 'semi-subsistence farming'.

If this case be accepted, there still remains the problem of defining the upper limit of sales that might be included in these classes. The difficulties of relating small cash sales to the produce consumed at home are formidable. The valuation of the domestic consumption in monetary terms, or the assessment of total production and sales in terms of starch equivalent, or some other material measure, is theoretically attractive. In both approaches variation of quality of produce adds to the difficulty of obtaining sufficiently accurate figures. Perhaps preferable is the proportion of farming time devoted to products for sale. Where other figures are available they can be converted into this measure through the concept of standard man-days, calculated for particular regions and types of farming (see Chapter 9).

As a starting point it is suggested that 25 per cent of farming time devoted to products for sale might be taken as the borderline between quasi-subsistence and semi-subsistence farming. The farmer who is devoting more than a quarter of his time to commercial products has a vital stake in the market, and success or failure in that market may decide whether the family lives in

[1] Winter (1956), quoted by Clark and Haswell (1964), 4.

tolerable comfort or below whatever is a locally acceptable idea of the poverty line. As long, however, as he is not devoting more than half his time to produce for sale the term 'semi-subsistence' would seem to indicate fairly the nature of the farmer's interest in the market.

My proposed classification is then:

(1) Subsistence farming:
  (1a) Pure subsistence
  (1b) Subsistence plus earnings for taxes, etc.
(2) Quasi-subsistence farming: not more than 25 per cent of working time devoted to cash crops.
(3) Semi-subsistence farming: 25–50 per cent of working time devoted to cash crops.

Owing to the lack of statistical information it is impracticable to confer upon such terms a high degree of accuracy except where detailed surveys are undertaken. Nevertheless, it is suggested that use of these terms, with an indication as to the accuracy assumed in any particular case, would add a useful degree of precision to reporting on the myriad variations of what is loosely referred to as 'subsistence agriculture' at the present time. Use of this terminology would also help to destroy the over-simplification of the primary division of agricultural economies into 'subsistence' and 'commercial.' There is a gradual blend from one to another and this is better suggested by the division here proposed. It may be borne in mind that state and collective farming makes another division, for these are not wholly commercial in their objectives.

The suggested classification is one of social organisation and economic objective. Alone, the terms here discussed tell us nothing about the type of farming, which demands a separate classification. Whittlesey recognised six subsistence types in his classification of world agriculture outlined in Chapter 4.

Whittlesey's first type, *nomadic herding*, has its commercial counterpart in livestock ranching in his classification but many herders are now organised in collectives. Those remaining nomadic herders are probably mainly subsistence or quasi-subsistence in economic type. His second type, *shifting cultivation*, is mainly subsistence in my proposed classification, though some examples would be quasi-subsistence. *Rudimentary sedentary tillage* is probably quasi-subsistence as commonly as it is subsistence. The other three types classed by Whittlesey as subsistence, viz. *intensive subsistence tillage with rice dominant, intensive subsistence tillage without paddy rice* (i.e. wet rice) and *subsistence crop and stock farming* contain all three of my classes, but in the last one semi-subsistence and quasi-subsistence economies have probably nearly ousted the subsistence type.

In a world classification it seems best to use type of farming, or crop and livestock combinations, for the primary classification, and economic class — subsistence or otherwise — to qualify this classification. We may then modify Whittlesey's classes to suit more readily the changed and fluid conditions of the present time. For example, his class *intensive subsistence tillage with rice dominant* would no longer include the word 'subsistence' in the title. 'Intensive tillage with rice dominant' adequately identifies the type, and does not destroy its close association with the monsoonal lowlands of south and east Asia. Description can be pursued with attention to the variation in the type from subsistence to semi-subsistence, and on to collective farms and any commercial units that may be concerned with this fundamentally unified type of farming.

Similar comments apply to the similar type without wet rice. It may be noted in passing that Whittlesey's definitions left a gap between the types 'with rice dominant' and 'without paddy rice'. It may be desirable here to distinguish intensive tillage with wet rice subsidiary, and intensive tillage with wet rice absent or unimportant.

The subsistence crop and stock farming type could disappear as noted in Chapter 4, regrouped into other types, ranging from rarely subsistence to commonly commercial and collective types. The other three 'subsistence' types did not include this term in their titles and so may remain unaltered, at least as far as this modification is concerned. *Nomadic herding* has undergone change of a drastic kind in some regions through collectivisation, but there are still large numbers of nomads whose economy is subsistence or quasi-subsistence. Nomadic herding in its more sedentary variations merges with and may ultimately become grouped with *rudimentary sedentary tillage*, modified in title to take account of the important, probably still dominant, livestock interest. The title 'rudimentary sedentary tillage' without emphasis on livestock remains applicable to large numbers of cultivators whose arrival at this type of agriculture has not been via nomadic herding, but rather through shifting agriculture. Here again, then, there is marked blending, but shifting cultivators and those who have become more or less sedentary but still retain some of the ways of the shifting cultivator offer the clearest examples of true subsistence agriculture. Even so, many shifting cultivators already have growing interests in cash crops.

Shifting Cultivation

As an example of agriculture which regionally is still of the subsistence class, though efforts have been made to transform it into what I am calling quasi-subsistence, a form of shifting cultivation will be examined. However, we are now confronted by another problem in definition to which reference must be made, that of the term 'shifting cultivation'. Various terms are used to describe this form of agriculture which is widespread in the tropical regions of the world. The essentials of the technique are the clearing of a patch of primary or secondary forest with axe or cutlass and fire, the planting of crops for a few years in the clearing, and movement to a new site when the fertility of the soil is seriously reduced. After a period of rest or fallow, the area will in due course be cleared again. These general principles suggest some of the alternative names that have been conferred on the practice, not only 'shifting cultivation', but 'slash and burn' and 'bush fallow' agriculture. Local terms include *ladang* (Indonesia), *caingin* (Philippines), *milpa* or *coamile* (Mexico), *ray* (Laos), *conuco* (Venezuela), *roça* (Brazil) and *masole* (lower Congo).[1]

The UNESCO Commission on World Land Use Survey[2] suggested that the term *shifting cultivation* should be used only when settlements as well as fields are moved frequently. They suggested *land rotation* where villages are fixed and the lands around farmed in a somewhat erratic rotation. The use of the word 'rotation' appears to the present writer to risk confusion

---

[1] Examples of terms listed by Gourou (1966), 31.
[2] UNESCO, *Report of the commission on world land-use survey* (1952).

with farming systems in which systematic rotation is practised.[1] A more immediate difficulty in the UNESCO terminology is pointed out by Nye and Greenland[2]:

> In fact, the two practices merge: thus among forest tribes, the Boro of the Amazon basin shift their settlements every few years; the people of Sa'a in Melanesia move when buildings become dilapidated, the cultivated land too remote, or if death and other misfortunes seem to dog the place;[3] the Ashanti in Ghana live in towns or villages that have endured for generations or even centuries. Recognition of a clear division between nomadic, semi-permanent, or permanent cultivators is also blurred by changes of population within a settlement. A family or clan will frequently move to found a new village or to expand an existing one. Nevertheless, so far as the soil is concerned what matters in all these cases is that the fertility of the land is improved by the natural regrowth of vegetation that springs up when cultivation 'shifts' to a new patch. . . .

Nye and Greenland make it clear that it is this process of fertility with which they are concerned, and they therefore use the term 'shifting cultivation' to describe cultivation practices rather than patterns of settlement. It is, however, obviously undesirable to use a term based on cultivation to describe a settlement pattern and therefore for this reason and that mentioned above, the term 'shifting cultivation' will here retain its general meaning, as interpreted by Nye and Greenland and other authorities. If further precision is required in terms of settlement the obvious distinction would seem to be 'shifting cultivation without movement of settlement' and 'shifting cultivation with movement of settlement' or as appropriate for intermediate forms.

To an observer unacquainted with its detailed composition and history an area of shifting cultivation presents a chaotic appearance:

[1] In such cases the usual term *crop rotation* is not wholly satisfactory, because it is not only crops that are rotated but complex systems of crop, grass and animal husbandry.

[2] Nye and Greenland (1960), 5–6.

[3] Forde (1934).

The first sight of native subsistence farming in the semi-deciduous forest region, e.g. in Ghana, presents an appearance of bewildering confusion to any one familiar only with the pattern of well ordered fields under single crops characteristic of more advanced systems of farming. There are no clear boundaries, individual fields can scarcely be discerned, and while some patches of land are definitely under crops, and others are under a thick regrowth of forest, there is a middle group in which perennial crops survive amidst a regrowth of forest which is gradually choking them. Some patches of land carry only one kind of crop, yet others appear to carry a mixture of up to half-a-dozen kinds in a seemingly haphazard arrangement. A pattern emerges from this higgledy-piggledy confusion if one follows the history of a single plot of land over a number of years.[1]

In the example to which we may now proceed for illustration of subsistence and quasi-subsistence agriculture, de Schlippe's valuable study of the Azande of the southern Sudan,[2] apparent confusion is again seen by patient study to be underlain by order:

When one enters a Zande homestead for the first time, the impression is that of complete chaos. The courtyard is shapeless or roughly circular or oval. . . . Crops, food and household belongings may lie about the courtyard, or be piled on to the veranda of a hut in what seems to be a most disorderly fashion. Worst of all, no fields can be seen. The thickets of plants surrounding the homestead seem as patchy and purposeless as any wild vegetation. It is impossible to distinguish a crop from a weed. It seems altogether incredible that a human intelligence should be responsible for this tangle.[3]

The tangle extends to family and group relationships, obligations and duties, and rights of ownership, including the ownership of land and crops. Individual ownership of land gives way, as commonly in such communities, to occupation of such areas as may be indicated from time to time by the tribal chief. The individual, however, owns the crop. Even this is complicated

[1] Nye and Greenland (1960), 1–2.
[2] Schlippe (1956).   [3] Schlippe (1956), 101.

Man and wife have fields of their own. The family is fed mainly from the fields belonging to the woman, so these crops go into her granary. The crops from the husband's fields go into his granary and are used mainly for entertaining guests. For most fields the wife has the main responsibility, but if she wants to sell any of these crops she must have her husband's permission and he decides on the distribution of the profit, but the husband has complete jurisdiction over his own crops. A second or third wife does not usually help the first in the fields, but daughters and small children help their mothers. Adolescent boys have small fields of their own for crops which do not require processing.[1]

Zande crops were found to be grown in associations of certain crops, sometimes sown simultaneously, sometimes successively. Each 'field type', as these associations were called, requires a particular ecological background.[2] A former refuse heap provides a specially favourable site for bananas, coco-yams, maize and eleusine (finger millet), the overhanging thatch of a kitchen roof shelters a tobacco nursery and hashish is sown in some corner of a field, where weeds help to conceal it from official eyes.

De Schlippe stresses the importance of the status of the woman as an economic 'atom' possessing her own equipment and field types, which he suggests may be a key to the understanding of an African system of agriculture.[3] 'In the Zande system of agriculture, and, I believe, in many similar systems at subsistence level, each smallest economic unit, in this case the woman, possesses a complete set of fields belonging to definite types in the same way in which it possesses complete sets of definite types of pots, mats, tools and buildings' (Figure 10).

A summary of a few of the field types reported by de Schlippe[4] will indicate the system underlying the apparent confusion:

(1) The *öti-moru*, or *main eleusine association*, may be established either on virgin land or in a second shift, that is on land which has been cultivated the year before. Work on this field type begins with clearing (May to July) and hoeing, followed, some twenty days later, by burning. This is called the 'hoe and burn' opening

---

[1] Schlippe (1956), 104–105.
[2] Schlippe (1956), ch. 3.
[3] Schlippe (1956), 106.
[4] Schlippe (1956), ch. 9, my summary. These are 'field types' of the 'green belt', exceptionally rich lands.

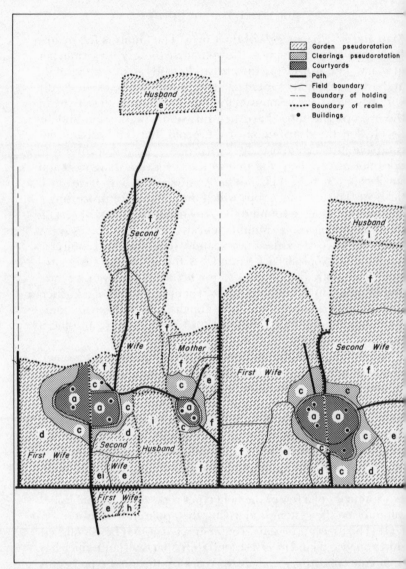

Figure 10. De Schlippe's map of adjacent holdings in the eastern green belt of the Z̄
country, showing the distribution of fields and realms. On the left, Kaimbaḡ
bigamous family, with a courtyard divided into two by a hedge of cassava and
a separate courtyard for the mother. On the right, a bigamous homestead w
courtyard which is not visibly divided into two realms. Key to field types: a, court
c, maize through sweet potatoes association; d, maize and oil-seed gourd associ
e, groundnut-eleusine succession; f, main eleusine association; h, bean patch; i, c
field. The ei field is an experimental association of groundnuts and cotton.

method. In grassless forest, hoeing is replaced by slashing of undergrowth with a machete. This is the 'slash and burn' opening method typical of the whole equatorial forest belt. On the very day of the burning and hand cleaning, or very soon after it, maize is sown in widely spaced holes and cassava cuttings may be planted. Two to seven weeks later the mixture of seeds of the eleusine association is broadcast and hoed in. Sesame and sorghums are commonly sown with the eleusine, less common associates being deccan hemp, oil-seed, water melons and cucumbers. A single weeding operation is carried out two to three months after broadcasting, one to two months before harvesting. The different crops become available for harvesting from September to January.

The main eleusine association can be established far from the homestead. It needs no guarding, only the maize and sorghum being liable to destruction.

(2) The *baawande*, 'place of groundnuts', is referred to as the *groundnut-eleusine succession*. With variations, the general procedure is similar, the hoe and burn or slash and burn opening being earlier, however, than in the previous case. Sowing of groundnuts follows at once (early April to late June), with maize or cassava as associate crops. This is the only big field type which needs constant guarding, jackals, guinea fowl, squirrels, pigs and monkeys all offering hazards at one time or another. A small guard hut is built in the main field and a man sleeps out armed with spear and grass torches. Groundnuts are the first crop available to end the seasonal food shortage, harvesting beginning late in June. Eleusine is sown as the groundnuts are harvested through July, August and September.

The broadcasting of eleusine through groundnuts utilises efficiently labour and the season. It also reduces to a minimum the period in which the soil is exposed to the heavy rains.

(3) The *bamvuo* or *eleusine through grass* association utilises the third shift or semi-exhausted land of a recent grass fallow. In the main eleusine association the period of rest from burning to broadcasting is meant to establish a low cover of grass through which the eleusine is broadcast and which is destroyed by the hoeing. Pre-sowing cultivation is unnecessary in the *bamvuo* but a thorough hand cleaning follows germination as in the main eleusine association. Sesame is the most usual associate crop and

may be the more important crop or even the only one, replacing eleusine in this field type.

(4) *Beans.* Patches of beans are very similar to the preceding field type. The seeds are sown through grass without pre-sowing cultivation and in this case hand cleaning is not needed. No guarding is needed so beans can be grown at a distance from the homestead.

The preceding four types are all independent of the homestead, being established in rapidly shifting clearings. The first two make use of land either newly cleaned or cropped the previous year, with appropriate variations. Type 3 is associated with a second phase and type 4 with second or subsequent phases.

(5) *Ridge cultivation.* This is associated with the homestead. An encircling ridge of earth is made up in February and March, maize and pumpkins being sown along the crest. The pumpkin vines creep down the inward slope into the courtyard and in June and July cuttings of sweet potato vines are planted on the outward slopes. This field type, which produces the earliest crops along with groundnuts, is the only one on which manures are applied, consciously by applying chaff and residues from salt preparation, and unconsciously through refuse sweepings. Other ridges are also made in the courtyard for yams, vegetables, rice and bananas. With this category go the old refuse heaps.

Other garden types typically include a strip for sweet potatoes on the outside of the ridge, often with maize and oil-seed gourds extending down the ridge, and plantings of cassava. Cassava provides a reserve against famine but if other foods were available it would be left unharvested.

De Schlippe's codification has been criticised as oversystematising Zande agricultural practices.[1] However, as summarised above, eliminating much of the detail, it illustrates the necessity recognised in traditional systems, of producing a variety of crops to cater for more than bare subsistence needs. Each field type occupies land at a particular stage, or one of two stages of the 'rotation'. Likewise each has its particular ecological background, higher or lower steps on the catena, transition forest or built-up ridge. Other special ecological situations are used often for specialised crops not included in the main associations, for example, the previously mentioned refuse heap and thatch

[1] Reining (1966), 73.

overhang. Others are ash accumulations, living trees for climbing yams, and termite mounds for sorghum, rice and cowpeas.

Cotton and 'The Zande Scheme'. All the field types so far mentioned have been concerned with subsistence crops, even if a little surplus might be sold from time to time. The introduction of a cash crop into the Zande system of agriculture followed the belief that economic development was necessary for progress, and that social progress of 'remote areas' could not indefinitely be financed by the economic development of the coastal belt.[1] The Zande district was selected for an experiment because it has the most reliable rainfall and the richest vegetation in the southern Sudan, together with the most disciplined population. Resettlement of the Azande people had already been undertaken in the 'twenties to facilitate control of sleeping sickness.

The new project acquired the name of 'The Zande Scheme' and it aimed at the complete social emergence and economic stability of the people. The central project was to be a cotton industry, providing for growing, spinning and weaving in the district and an industrial centre was established between 1946 and 1949. Also during these years resettlement was undertaken and 60,000 cultivators were given new, dispersed, holdings. It was hoped that this would be the beginning of a transition from shifting to fixed cultivation, and in any case would facilitate supervision of the cotton crop. Each Zande cultivator received a holding some 800 to 1000 metres long and 150 metres wide, with a frontage on a common path. Fifty to sixty holdings formed a settlement under one elder. The Azande were allowed to erect their homesteads at any point within their holdings.[2] A homestead is not moved without good cause, such as marriage, death of a wife, repeated family misfortunes or persistent crop failures.[3] The early years, particularly the first, in a new homestead are difficult because of all the extra work to be done.

It soon became evident that the incorporation of a demanding cash crop into the Zande scheme of cultivation presented many difficulties. Sowing was often delayed because the Azande were reluctant to abandon all other activities to concentrate on cotton. Much weeding was required, in contrast to the customary field types which require none, or only one weeding. Picking, which should begin 140 days after sowing and be finished some 70 days

[1] Schlippe (1956), 20. [2] Schlippe (1956), 21. [3] Schlippe (1956), 192.

later, was also often delayed and was very inefficient. Cotton clashed with the customary field types in labour utilisation, especially the main eleusine association.[1]

Before the British administration withdrew from the Sudan it had become clear that the scheme had failed in its main objective to provide a stable form of cash cropping which would induce the cultivators to adopt a progressive and sedentary way of life. The administrators failed to gain the confidence of the people in spite of earnest and sincere intentions to improve their lot through wise and profitable cultivation. The trouble lay largely in the failure of the Europeans to understand the Zande outlook, and in a general lack of communication, while the low prices paid for the cotton, in an attempt to moderate change and guarantee stability, led to disillusionment on the part of the growers.[2]

The Zande scheme illustrates the risks inherent in interference with a system of agriculture which has grown up as part of a complete culture and a set of values quite different from those of a commercial and industrial society, however well-meant the scheme devised. Only by patient enquiry can a system be thoroughly understood and even then to prescribe remedies for the undoubted evils of an existing system may still be fraught with peril. Long-term education so that the indigenous people may be able to be largely responsible for their own approaches to development — an acceleration of the process of change that affects all peoples in some way — may be the only answer.

Even education is not always wholly beneficial. If it leads to the break-up of tribal life and abandonment of traditional values it may cause increased tension in the society and the downgrading of agriculture in the people's estimation. This inevitably leads to increased migration to the towns which, in turn, can offer to many only unemployment and deepening poverty. The urbanised, landless migrants have lost the stability and dignity of their tribal life and there are more mouths to be fed by others.

Apart from the political and social questions associated with primitive forms of agriculture there are uncertainties about their record in the conservation of fertility and other resources. Shifting agriculture, in particular, was formerly much attacked,

[1] Schlippe (1956), 137–138.
[2] Reining (1966), 220–232.

as in the pronouncement by the U.N. Food and Agricultural Organisation that:

> Shifting cultivation in the humid tropical countries is the greatest obstacle not only to the immediate increase of agricultural production, but also to the conservation of the production potential of the future, in the form of soils and forests.[1]

Opposing views based on the past success of shifting cultivation in sustaining growing populations and the dangers of disrupting tribal life and organisation have received some support from the technical investigations of soil scientists. Nye and Greenland show that the question of whether or not shifting cultivation leads to a squandering of resources of the land admits of no simple answer.[2] Much depends on the particular environment, there being far less to be said in favour of the system in savanna than in forest regions. Improvement of the system may be possible, but there is no evidence yet that a planted fallow will restore the fertility of a forest soil any faster than the natural woody fallow. In particular, the nitrogen status maintained beneath a forest fallow is good. In other respects much depends on variations in the system. Useful elements in the fallow may be employed, such as bananas, a semi-perennial crop, while in savanna regions livestock may provide an answer. Greater availability of cheap fertilisers may be the key to evolution of farming systems here, as in temperate latitudes.

Again one is reminded of the variety to be found within a seemingly simple label. Watters has suggested a classification[3] based on predominant and subsidiary ways of producing food by combinations of shifting agriculture, permanent cultivation, pastoralism, hunting, fishing and gathering. The potential for development will vary enormously from one type to another as well as from one region to another. What is common to all types and all regions, with insignificant exceptions, and forces changes in the agricultural system, is the growth of population at rates much more rapid than in the past. Even the modest aids to health and longer life-span that reach the communities of the forest and savannah are having this effect.

[1] FAO staff, Shifting cultivation. *Unasylva.* 11, No. 1, 9–11, reprinted in *Trop., Agric. Trin. 34*, 159–164.

[2] Nye and Greenland (1960), esp. ch. 8.

[3] Watters (1960).

Gourou[1] quoted studies indicating 3–12 persons per square km as typical, with lower optimum figures in many cases. In contrast, a population of 500 per square km was estimated for the alluvial rice lands of the Tonkin delta of Vietnam, rising to 1500 per square km in some parts. In Tonga, bush fallowing has survived increases in population to over 200 per square km but fallow periods have, of necessity, been reduced, and crops introduced into fallow land.[2]

One cannot, of course, expect that lands used for shifting cultivation will necessarily have a very high potential under any other system, so that the above contrast may be regarded as illustrating inherent land potential as well as the productivity of a system of agriculture. Land in shifting cultivation may, however, be suited to a variety of uses, including plantation agriculture, and the alienation of tribal lands for this purpose or for resettlement of other tribes has been the cause of privation for some communities. It was not always in the past realised how large an area might be needed for the maintenance of a family, and that land found apparently unoccupied might be part of their bush fallow though miles from the homestead.

On the other hand, formation of plantations or the introduction of settlers of more advanced techniques may ultimately have beneficial effects on a primitive tribe. Similarly, sedentary cultivators may be affected for either good or ill by the establishment of plantations in their locality. An initial impact is the attraction of the plantation in providing regular employment. This may act as a disrupting force to local cultivation, but if there is surplus labour it may, on balance, prove economically beneficial.

Peasant Farming in Malaysia

Prior to the colonial era the culture of the Malays was based on rice supported by fruits, coconuts, vegetables and spices. This truly subsistence economy was disturbed by British rule, or British protection of the native ruler, causing the spread of a monetary economy, largely through the introduction of the plantation system. Fortunately, the requirements of rubber led to the development of the rolling, well-drained lands, with much clearance of forest. The low, wet lands suitable for rice were left in

[1] Gourou (1966), 47, 119.   [2] Maude (1970), 60–62.

the hands of the indigenous farmers and legislative control over the transfer of land helped to maintain this position. Moreover, the Malays were largely indifferent to the demands of the plantations for labour, preferring their independent, rather casual farming to the regular and monotonous plantation employment. Hence, the labour force became based mainly on Indian and Chinese immigrants. But the growing markets for crops and more stable conditions, particularly the suppression of banditry, made it more attractive to grow surplus crops for sale. The growth of the rubber industry provided outlets for peasant as well as plantation. The enthusiasm with which Malay peasants turned to rubber growing[1] is a measure of the attraction of a commercial economy, in spite of the small size of farms. It might be expected that unless a holding amounted to 8 hectares or so, it would be used for intensive crop production. In fact, even in Johore, where 94 per cent of the peasant holdings were under 6 hectares in size, rice cultivation became unimportant compared with cash cropping based on rubber, coconut and pineapple.[2] In 1974, rubber occupied 1·7 million hectares, 52 per cent of the total cultivated area, and two-thirds of the rubber was on smallholdings.[3] Crops other than rubber, coconuts, pineapples and bananas are grown by Malays mainly for domestic consumption.

The main areas of paddy growing are the northern coastal plains and it is estimated that the growers in Perlis and Kedah sell about half their crop, while in other coastal plain areas paddy farmers have about one-third of their crop for sale. During the annual fallow, selected areas of the paddy fields are planted with maize, groundnuts and vegetables mainly for domestic consumption. Inland, small paddy fields occur widely in the valley bottoms with tree crops on the interfluves (Figure 11). In these areas paddy is a subsistence crop while cash is obtained from rubber and some of the coconuts and other fruit grown.

Even the Malays of remote jungle areas purchase some goods from the proceeds of nipah palm leaves (for thatch) rattan (for baskets, mats and furniture) gums, resins and other forest products. Hence, it can be said that an entirely subsistence

---

[1] Ooi Jin-bee (1963), 202, 204.
[2] Ooi Jin-bee (1963), 193.
[3] Voon Phin Keong (1972) provides a useful case study.

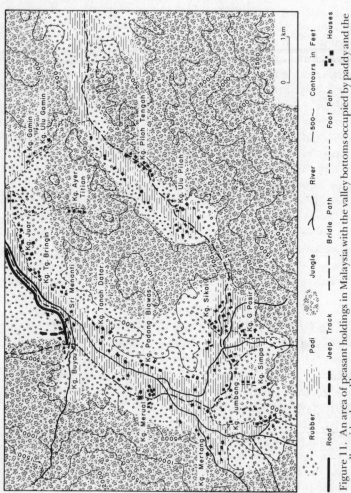

Figure 11. An area of peasant holdings in Malaysia with the valley bottoms occupied by paddy and the valley sides by rubber.

Source: Ooi Jin-bee, *The Journal of Tropical Geography*, Vol. 15, 1961.

economy does not exist today among the Malays.[1] Most would be, in terms of the classification proposed above, quasi- and semi-subsistence rather than subsistence farmers.

If the Malay is firmly attached to the money economy, this is even more true of the Chinese peasant in Malaysia. The Chinese, who migrated to the Malay peninsula in large numbers after the establishment of British rule, sought occupations which would enable them to make profits and eventually return to China. They therefore grew cash crops in high demand, notably vegetables in intensive market gardening. No doubt, the success of the Chinese immigrants encouraged the Malays to pay more attention to producing for sale, but they have never rivalled the Chinese in intensive farming. One reason is that the Muslim objection to the pig prevents the Malays from integrating pig and crop farming as the Chinese do, using pig manure for the crops and crop residues for the pigs. Indian and Pakistani peasants are far fewer in number but are also engaged mainly in production of cash crops, notably rubber.

From the national point of view the preference of the smallholder for cash crop production has some unfortunate results. The demand for rice is not satisfied by home production and large sums of hard-earned foreign exchange have to be spent on imports. But from the point of view of the individual smallholder it makes good sense. The wet and cloudy climate does not greatly favour rice growing and the smallholder can normally buy more rice with the proceeds of an acre of rubber than he can grow on the same area.

For examples of true subsistence economies in the Malay peninsula it is necessary to turn to the aboriginal population. This comprises Senoi, Aboriginal Malay, Negrito and many lesser groups, totalling in all perhaps 100,000 people. The numbers of such people in the Indonesian islands and New Guinea are much greater. Many of these tribes have not changed their way of life during the thousands of years since man first entered the region and they remain virtually in the stone age. Since hunting and gathering are not regarded in this book as agriculture, little space will be devoted to comment on the economies of these people, who use spear, trap, and blow-pipe to add to the berries, roots and fruit collected in the jungles. It may be remarked, however, that

[1] Ooi Jin-bee (1963), 163.

even these groups are being slowly affected by contact with more advanced cultures and may plant a few food crops in places which are convenient for them to return to for such harvest as may survive. In Malaysia, shifting cultivation is practised by many of these folk and some establish rubber trees, which, with the proceeds of collected forest products, enables them to make a few purchases. As many as, possibly, 15 per cent of these aboriginals in Malaysia now live a settled life, cultivating paddy and rubber.

PART III
CONCEPTS AND METHODOLOGY

CHAPTER 9

Systematic and Regional Analysis

The first part of this book provided a general background of basic material on the physical and human environmental factors which particularly influence the pursuit of agriculture. The second showed how in different areas completely different responses to the overall necessity to produce food and raw materials from the soil have evolved, and in these contrasts can be seen the influences of factors previously discussed in general terms. We note emergence of different types of farming, described by reference to the crops and livestock produced, of different approaches to the organisational problems based on relatively small private farms, or large plantations and collectives, and of regional patterns of these variations. In the course of description, problems both practical and academic are revealed, some of which are very relevant to the global problems of food supply which were referred to in the introduction to the book. This chapter and the next show some of the ways in which these problems can be subjected to analysis.

The descriptions in the preceding chapters of selected types of farming, their organisation and the regions in which they are practised, result from the synthesis of many studies, building up the picture of the whole. Before the synthesis can take place, data must be collected, sifted and analysed. This progress should be followed for each investigation. There is a tendency in scientific circles today to think or to imply that quantitative analysis is more academically respectable and more intellectual than the non-quantitative, which with rather pejorative overtones is called 'descriptive'. In fact, valuable analysis of a non-quantitative kind is possible, and is indeed constantly pursued in all walks of life from abstract philosophy down to the daily tasks of the individual. Also, it is necessary to remember that quantitative work is itself descriptive — the description being through figures rather more than through words. Quantitative analysis moreover, has to be interpreted when completed, and is often wrongly interpreted, partly because there are no foolproof methods of arranging and analysing any set of data. The best results are still likely to be produced by a thoughtful consideration of the widest possible range of information bearing on the problem to be solved, the quantitative data being subjected to appropriate mathematical processing, the non-quantitative being treated to critical attention, and conclusions drawn from both kinds of material.

The essential virtue of quantitative data is that it can be classified and replicated with precision, so that comparison over time and space is possible with a minimum of ambiguity. However, the results can be no better than the raw material and much data are suspect, so caution is always needed. Definition of the data and comparability over time and space must be sought carefully, and the errors and weaknesses of different methods of classification must be recognised.

It is with the analysis of quantitative material that we shall be largely concerned in this chapter. Whether studying a type of crop farming or livestock production world-wide, the mixture of types of farming in a region, or the delimitation of a region, the investigator will wish to use as much numerical data as is available and reliable.

The data used in any investigation will depend upon the objective of the research and the available material. Collection of data for individual research projects is time-consuming and expensive but may be essential in the case of projects covering small areas, for which statistics are not available from official or other sources, when investigating at the farm level, for which statistics, even if available in official files, may not be divulged to researchers, or if the line of enquiry concerns a matter for which data have not previously been collected. In investigations covering a large area, however, it is generally necessary to use statistics collected in official census enquiries and published or made available in aggregate form to research workers. A detailed census of agricultural holdings and activities is made regularly in most developed countries[1] and is becoming increasingly common in the less developed countries, though collection of data may

---

[1] Tarrant (1974) summarises the United Kingdom, U.S.A. and Swedish agricultural censuses and considers problems arising in their use.

here be fraught with many difficulties and the results correspond-ingly of doubtful accuracy.

The value of investigating systematically problems such as low productivity, poverty, incidence of pests and diseases, use of and response to fertilisers, variation in labour supply, input and output relationships and amalgamation or division of farm holdings is self-evident. They are problems that can be seen to have a direct bearing on the availability of food supplies and the prices at which these supplies can be marketed in the future as well as in the past and the present. Few investigations, however important, can be carried out in great detail on a world-wide basis, but if scientifically designed sampling is undertaken the results *may* have world-wide validity. National or regional studies are more common, and sampling is often necessary in these cases also in order to keep an enquiry within the bounds of practicabil-ity in time and cost.

Of less obvious practical value is the research into the methods of delimitation of regions which has occupied an immense amount of time in geographical studies. The call for recognition of regions is, however, widespread, for planning and other practical reasons. It was implied above in the reference to 'national or regional studies' that an enquiry may often profitably be carried out in an area other than that of a whole national state. The area may be smaller, as in the case of one part of a country that has particular problems — for example, because only in that part is a particular crop grown — or because its farms are particularly prone to poverty or marginal conditions, or it may be larger than one state, extending across international boundaries. Rarely do existing administrative boundaries effectively indicate the area that needs to be examined and the question of delimitation involves the application of principles of regionalisa-tion before the study in depth of the problem can commence.

A region may be defined as a part of the planet's surface having certain characteristics which enable it to be recognised as a unit, distinguishable from other units which surround it, and which may themselves be identified by similar or different characteristic features. A region may be defined in physical terms, e.g. a volcanic plateau, a river valley; in climatic terms, e.g. a tropical monsoon region; or by economic and social criteria, such as a textile-manufacturing region, or a nomadic herding region. A single point in space may thus be included in several regions, differently defined, the boundaries of which will rarely coincide over much of their length, except where some marked physical feature such as a seaboard or steep mountain wall sets a limit.

It is partly because of this variability of boundaries that the methods by which regions may be defined have been examined and re-examined repeatedly during the past half-century, and ever more complex procedures evolved. Not only will the limits of a region differ according to the criteria by which it is defined but, at least in some cases, these limits will change with time.

Boundary lines are in themselves unsatisfactory. Most regional variations occur in a more or less gradual way, so that zones rather than lines should be used as boundaries. On a small-scale map, of course, even a fine line does, in fact, represent a broad zone, though it may still be too narrow to suggest accurately the real width of the transitional area. Provided, however, we remember that boundary lines can be interpreted only as more or less accurate indications of zones within which marked changes in the regional pattern occur, a map of regions can summarise knowledge of spatial variations without seriously misleading the careful interpreter.

During the evolution of geography as an academic and practical discipline, the concept of regionalisation played an important part. It is not long since the task of recognising and delimiting regions was assumed to be a central task in geography,[1] and though it was appreciated that the analytical tools available were inadequate it was generally believed that gradual improvement of techniques would lead eventually to the solution of most, if not all, the problems inherent in delimiting regions. More recently, however, the regional approach has been widely discredited as geographers have sought to escape from the limitations of 'descriptive' studies and to base their research work on sounder scientific principles. McCarty and Lindberg see little analytical value in regionalisation[2] and Anuchin considers it has no value as a goal.[3] Even these critics, however, admit to some uses for regionalisation in descriptive studies, which, presumably, may precede more analytical treatment.

[1] See, for example, Hartshorne (1939, 1959), Bunge (1962), Grigg (1965, 1969).
[2] McCarty and Lindberg (1966), 95 ff, 103 ff.
[3] Anuchin (1961), 34.

While geographers have doubted their ability to perform the task and even the underlying justification for it, the demands for regionalisation have grown immensely in the fields of planning, and government activity generally, as noted above, and the question may fairly be put as to what discipline, if not geography, should conduct research into what is possible in regionalisation, and what errors should be avoided. There seems little doubt that, whatever academic view geographers adopt, the practical demand for delimitation of regions as spatial frameworks for policy application will continue and there would seem to be much to be said for the view that geographers should make available their expertise in spatial description and analysis, while clearly indicating their understanding of the pitfalls.

Spencer and Horvath remark that agricultural regions are 'expressions of the subjective choices of man operating in groups, affected by a myriad of cultural influences, all produced by man himself'. Experience leads to many individuals making similar selections so that, in effect, the operation becomes a group procedure. They suggest that herein 'lies the determination of a way of life and the patterns of crop combinations which the geographer can recognise as regional expressions'. They identify six different categories of cultural processes significant to the cultural origin, maturity and change of the agricultural region. These are psychological, including acceptance of new ideas and processes of diffusion, political, historical, technological, economic and agronomic or management processes.[1]

The description of a region and analysis of its problems should include consideration of all these processes of a cultural nature as well as, of course, the physical environment, but first the region must be recognised and decisions taken on the type of regionalisation study to be pursued.

Gregor[2] recognises three distinctive types of agricultural regionalisation, single-feature regions, such as the areas occupied by a single crop; multiple-feature regions, such as land-capability and farming-system regions; and total agricultural regions as attempted by Hahn in the nineteenth century[3] and Whittlesey in the 1930s.[4] He notes further that the comparison of regions is 'for many the crowning step in regional studies'.[5] Though differently

---

[1] Spencer and Horvath (1963).　　[3] Hahn (1892).　　[5] Gregor (1970), 125.
[2] Gregor (1970).　　[4] Whittlesey (1936).

ordered, there is consideration of each of Gregor's three types of region in this book. Whittlesey's regions have already been described and briefly discussed.

Buchanan has pointed out[1] what may seem obvious but had certainly not been observed in all previous practice[2] that agricultural regions should be defined in agricultural terms — by means of a crop, a crop association, or even a system of organisation of farm processes. Such criteria are the growing of spring wheat as the indicator of the Spring Wheat Belt of the United States and Canadian prairie provinces, and the association of wheat, alfalfa and cattle for the Wheat-Alfalfa-Cattle Crescent of the Argentine pampas. Some such regions are readily recognisable on the ground even to the untrained eye, at least in core areas. Two such areas long recognised in the United States are the Corn Belt and the Cotton Belt. When we come to fix boundaries to these regions, however, we soon find need to clarify observations by resort to theory supported by mapping and statistical analysis. Thus, we may decide to define a particular crop region as the area in which a certain percentage of farms have not less than a specified percentage of their area in that crop. We must remember, however, that other products may be even more important on some farms in the 'belt' than that which is dominant in the region, and one belt may overlap with another. Furthermore, an area 'simply occupied by certain features and defined by certain arbitrarily chosen criteria' is frequently interpreted as an economic region,[3] implying far greater importance and refinement of classification than is justified by the distinctiveness of the parameter, and we must be careful to avoid such errors.

This and the following chapter are not, however, planned primarily to justify or even to illustrate regionalisation as such, but to set out some of the basic approaches to the study of distributions of agricultural phenomena, whether or not they are to be grouped into regions. Essentially, they are concerned with methods of distinguishing 'what is located where and why'.

[1] Buchanan (1959), 5.
[2] For example, Whittlesey used the term 'Mediterranean agriculture', following earlier practice.
[3] Morgan and Munton (1971), 127, 138.

Theories of the Location of Agriculture

An attempt to offer a theory which could explain the location of different types of agriculture according to economic principles was made by Johann von Thünen in the first half of the nineteenth century.[1] Von Thünen, who managed an agricultural estate in Mecklenburg, near the city of Rostock, observed the disposition of different agricultural activities in his area and related these to the farm accounts he managed. The theory he produced has provided the starting point for many discussions of regional location by economists and geographers. It must be realised at the outset that von Thünen's theory can be judged only in the restrictive conditions he postulated and in the historical conditions of his time. There are obvious discrepancies between the theory and observed fact, but to criticise the theory on this account, neglecting the postulations, is improper, as with any theory.[2] Essential to the theory were the postulations that surrounding one central city in an 'isolated state' was an area of uniform physical conditions suitable for tillage, and all surplus produce of the region had to be sold in the city. On the assumption, further, that all transport was by one means (horse-drawn) and transport costs were proportional to distance and borne by the farmer, then the result would be a series of concentric rings or belts of differing production around the city (Figure 12A). The boundaries of each belt would be defined by the principle of economic rent,[3] in effect the maximum profit per acre having regard to transport costs. In von Thünen's time, wood, in great demand in a city for fuel, could return higher profits per acre than rye, provided transport costs were low, but cost of marketing limited the zone within which this would apply. In fact, in von Thünen's model, the wood-producing zone was

---

[1] von Thünen (1826). For an edited translation, see Hall (1966).

[2] For sympathetic discussion of von Thünen's theory see Grotewald (1959), Chisholm (1962) and Tarrant (1974).

[3] In orthodox economic theory rent is the surplus or excess payment obtainable for a factor of production on account of its inherent qualities, e.g. land of special productivity or in a particularly desirable position. It is an unearned payment to the owner, which is not necessary to bring the land into production but which the user is willing to pay because yield is greater than on alternative land. The concept, propounded by Ricardo early in the nineteenth century, is discussed in all textbooks of economics and a clear description in relation to von Thünen's theories is given by Chisholm (1962), Chapter 2. Later in the nineteenth century Jevons and Marshall made the more exact proposition that rent should be determined by the marginal productivity of land.

nearest to the city excepting only for the zone of market
gardening and milk production, of which the location was fixed
by slow transport and consequent high prices for perishable
goods. Beyond the wood-producing zone were belts of arable
production, of decreasing intensity. Higher costs of transport
were offset by lower production costs. Beyond the third zone was
an area of the least intensive system, livestock farming.

Although the stipulated conditions of the isolated state and one
form of transport do not occur in practice, and competitive
marketing and variable transport costs, to say nothing of differing
soils, complicate the position in every area, some traces of the type

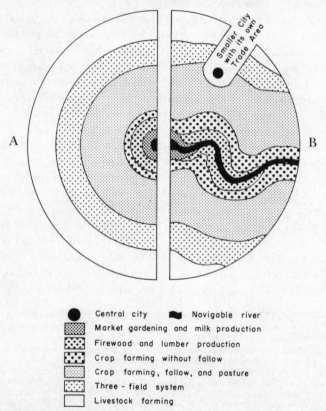

Figure 12. Von Thünen's agricultural zones. (A) The original isolated state, (B)
an example of modified conditions.

Source: A. Grotewald, *Economic Geography*, Vol. 35, 1959.

of distribution envisaged by von Thünen can still be found. A practical reason applicable in any circumstance is that it costs more to obtain goods from further afield or to cultivate land further from a farmstead or settlement than nearer, unless transport costs are outweighed by differences of soil fertility, climate, land prices or other variants. Except where mechanised transport is available, even short distances to be traversed to work in the fields greatly reduce the time and energy available for productive work. Hence, nearer fields are preferred for the more intensive forms of cultivation. Impressive zoning around agricultural villages has been observed in many countries and a number of examples are cited by Chisholm,[1] together with consideration of the background of costs and time expenditure.

An attempt to escape such constraints in field utilisation has been adopted in the U.S.S.R., where collective and state farms have been steadily increased in size, as seen in Chapter 7. The benefits of larger, centralised settlements are stressed as facilitating provision of modern services, with motor transport being used to overcome the separation of village and fields. Where the more moderately sized farms of private enterprise prevail, the changes from traditional patterns are less marked, but the growing tendency is for farms to be consolidated to bring economies of scale, the truck and tractor providing the means of reducing drastically the time lost in transport to and from the fields.

In subsistence economies where domestic refuse is used as fertiliser, the nearer fields obviously benefit from their location. Examples of this were given in Chapter 8, and Chisholm quotes others.[2] Siddle shows the value of a von Thünen model in studying the spacing of rural settlements in a subsistence economy (Sierra Leone) in which the economic radius for land use and kinship considerations govern village size.[3] However, distance variables will usually afford explanations of only some changes of pattern and, as Blaikie points out, testing in the field usually calls for explanation other than that offered by location theory. In an examination of two groups of Indian villages, he found that variations in levels of total output were undoubtedly, in one case, largely controlled by the mean manuring level and

[1] Chisholm (1962), Chapter 4.   [3] Siddle (1970).
[2] Chisholm (1962), Chapter 5.

this in turn was controlled principally by variations in the mean distance of fields from the cultivators' homes. In other villages, however, where irrigated crops, including cash crops, were grown, decisions on field use were dominated by the availability of water, either from wells or canals, and in the latter case proximity to the inlet gate and low level location were all important.[1] However, Blaikie found that whereas, at the farm level, locational considerations were generally residual and unimportant in cropping strategies, 'regional' patterns of crops, levels of total input and manuring, and irrigation facilities responded well to geographical analysis at the aggregated level.

In an early attempt to explain the regional variation in agricultural patterns, Jonasson, the Swedish geographer, prepared a diagram (reproduced as Figure 13), which in effect represented an adaptation of the von Thünen model to the agriculture of Europe in 1925.[2] The zones of production he postulated were:

I Horticulture.
    Zone 1. The city itself and immediate environs. Greenhouses, floriculture.
    Zone 2. Truck products, fruits, potatoes, and tobacco (and horses).

II Intensive agriculture with intensive dairying.
    Zone 3. Dairy products, beef cattle, sheep for mutton, veal, forage crops, oats, flax for fibre.
    Zone 4. General farming; grain, hay, livestock.

III Extensive agriculture.
    Zone 5. Bread cereals and flax for oil.

IV Extensive pasture.
    Zone 6. Cattle (beef and range), horses (range), and sheep (range), salt, smoked, refrigerated and canned meats, bones, tallow and hides.

V Forest culture.
    Zone 7. The outermost peripheral area. Forests.

---

[1] Blaikie (1971), Part II, 21. Blaikie's methods are briefly described on page 219 of this chapter.
[2] Jonasson (1925).

Jonasson recognised that an actual case of such effect of distance from market upon the gradation of land utilisation did not exist but cited examples which offered parallels. He quoted Moscow, Indianapolis and Buenos Aires in this context. He was evidently impressed by the finding of conditions not far from the ideal on the Edwards plateau in Texas,[1] and he reproduced these conditions as concentric circles with six zones showing similar variation from horticulture to ranching.

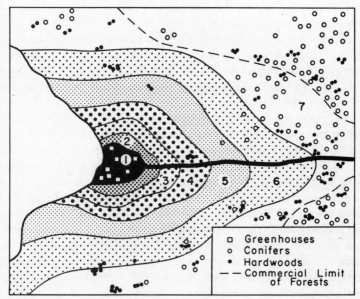

Figure 13. Model of the zones of production about a theoretical, isolated city in Europe. For key to zones see text.

The most regularly occurring element in this pattern of concentric agricultural regions is the zone of market gardening found adjacent to many cities. This survives because people are prepared to pay prices for fresh vegetables which yield often a higher return to the grower, given short hauls and therefore low transport costs, than is available from selling vegetables for freezing or canning. Thus, there is an incentive for growers to locate close to city markets. Furthermore, as noted in Chapter 3,

[1] Youngblood and Cox (1922), cited by Jonasson (1925), 285.

the return is sufficient to enable the market gardener to pay higher prices for purchase or rent of his land than could be sustained by other forms of agricultural production. Surrounding, or interwoven with, this zone is commonly also found specialisation in producing fresh milk for the urban areas. Although rapid transport enables cities to draw milk from great distances today, intensive dairying can return higher revenues than less intensive forms of farming and therefore can tolerate higher land values.

The combination of these forms of land use influenced by the interplay of economic and physical factors is well seen around Christchurch, a city which is relatively isolated from other large towns in the South Island of New Zealand. A zone of market gardening land straddles its boundaries but does not surround the city. Fat lamb production, the dominant activity on the adjacent Canterbury Plains, is carried on immediately to the west of the city, where soils are light and dry. The difference in different kinds of dairying enterprises is also marked. Town milk supply obtains much higher prices for milk than the alternative outlet provided by the butter or cheese factory. Hence, farmers who undertake dairy production near towns do so for town supply, whereas those more distant cannot compete in this and have to accept butterfat prices. Fat lamb production is more attractive to these farmers than milk production for manufacture and so is interwoven with town milk supply, and dominates the areas beyond. In the more distant hill country are farms producing store sheep. The low-intensity store sheep farming in the hill country is maintained there, however, not because it can withstand high transport costs, but because of the low physical productivity of the land, which necessitates extensive grazing by hardy livestock.

Thus, in this region there is some zoning in the pattern of agricultural land use reminiscent of that postulated by von Thünen, but soil and climatic factors in this case limit the play of transport costs. It is, of course, precisely the importance of transport costs that is illustrated by von Thünen's reasoning, but perhaps too much attention has been given by geographers to looking for occurrences of 'von Thünen patterns' in practice. The predictive element in von Thünen is very limited and continuous preoccupation with this theory, however much in advance of its time it may have been, has distracted attention from alternative and more advanced analysis.

The concept of economic rent is itself of only limited relevance in the practical world and if it is to be applied it must be related to marginal analysis. For each factor of production there is a marginal product as well as an average product and it is the marginal product which decides the level of cost that it is worth incurring, i.e. it is worth incurring a marginal cost rising to the point where it is equal to the marginal product (or increasing output until the marginal product falls to the level of the marginal cost as in the extension of cultivation to poorer or more remote areas). Beyond this level any increase in output will decrease profit. The significance of marginal productivity of land and other inputs may be assessed by calculating production functions. These are mathematical expressions showing the expected increment of product for varying inputs of land, labour, seed, fertiliser, etc. The main problem with production functions is the complexity of inputs needing to be examined and the desirability of using data for individual farms, though useful progress has been made with grouped data. The value of production functions applied to advanced systems of agriculture as well as simple systems has been demonstrated in many investigations. From them it can be seen whether marginal returns to inputs are larger or smaller than the marginal costs of the inputs and recommendations can be made to increase one input while reducing another to obtain optimum use of the land. Eventually such data may lead to significant advances in developing a useful theory of land use which could help to predict future changes with given variations in costs and prices.[1]

The search for a general theory of location of industry which has occupied many economists since von Thünen's day has been concerned mainly with the distribution of manufacturing industry rather than of agriculture. The theories of Weber,[2] Hoover,[3] Lösch[4] and others who have investigated the advantages to be derived by the manufacturing firm from locating at the point where its costs will be least and/or revenue will be at its maximum are, however, relevant to the study of the theoretical best location

[1] Clark (1973), Chapter 3.
[2] Weber (1909).
[3] Hoover (1948).
[4] Lösch (1940, 1954).

of agricultural enterprises. Isard[1] has attempted to integrate industrial and agricultural aspects of location through land use, and Dunn[2] has applied location theory specifically to agriculture.

A location theory should provide the answers to a number of questions. For the individual farmer there should be guidance on where he should locate his farm to produce given products, how large the farm should be and how intensively it should be cultivated. Regional and national authorities should be able to obtain solutions to problems of economic and regional planning. So far these objectives have not been achieved, in spite of growing complexity in the methods of economic analysis employed, which demand a considerable mathematical knowledge for their full appreciation. No model can yet cope with the complexity of background, the intricacy of human reasoning and the constant change that affects all economies. Dunn aims to provide, and to some extent succeeds in achieving 'a general framework of analysis that will serve the dual function of aiding in the evaluation of existing techniques in the field of agricultural economics, and suggesting new approaches to the study of agricultural phenomena.'[3] Understanding of the orientation of production and of the fact that the spatial orientation is an inseparable part of the total economic problem is carried a little further than previously. But it is not suggested that a model is offered which is capable of direct application to the solution of location problems. 'One cannot insert data into a machine and receive an answer sheet that defines the optimum geographic distribution of agricultural production. There are too many sources of discontinuity and elements of indeterminacy in the realistic case'.[4]

Crossley has adapted Dunn's rent theory to explain an actual case of agricultural land use in response to industrial processing for a world market.[5] The alternative products from beef cattle include fresh, chilled, frozen and corned beef and beef extract, with producers of cattle of successively lower quality and the processing plants located at increasing distances from the central market. The application of the marginal rent principle explains this location pattern. According to Dunn, 'that form of land use

[1] Isard (1956).
[2] Dunn (1954).
[3] Dunn (1954), 93.
[4] Dunn (1954), 93.
[5] Crossley (1976).

which provides the greatest rent will make the highest bid for the land and hence displace all others'.[1] Crossley rephrased this as 'that form of beef cattle use which provides the greatest rent will make the highest bid for the beef cattle and hence displace all others'.[2] Dunn's equation $R = E(p-a) - Efk$ is then redefined to let R equal the rent per unit of beef cattle, and the remaining terms in units of beef product; E equalling product per cattle unit, p the market price, a the production cost, f the transport rate per unit of distance and k the distance. Although Crossley based his study on 1923–25 when trade patterns were relatively little distorted by political consideration he claims that the model has a predictive value.

The development of predictive models is now becoming urgent both in terms of the maturity of geographical thought and practice and their possible contribution to the solutions of practical economic problems. So far their application has been mainly in helping empirical investigation proceed more critically.

Empirical Studies

As Grotewald[3] has pointed out, facts actually observed and generalisations derived from them are likely to be more valid than *a priori* theories, because they do not rest upon propositions and therefore their relevance to reality cannot be doubted. There are many differing methods of empirical analysis that deserve attention by the student who would understand what is meant by an agricultural region.

The pioneer work by O. E. Baker and his co-workers first calls for comment. O. E. Baker, economist in the U.S. Bureau of Agricultural Economics, stated the physical and economic principles which explain the distribution of crops in the first issue of the journal, *Economic Geography*[4]:

1. The crop or other agricultural product which is most limited in climatic or other physical requirements of production will, if the demand for it be sufficient, have first choice of the land. It possesses, so to speak, a sort of natural monopoly

[1] Dunn (1954), 6, footnote 5.
[2] Crossley (1976), 61.
[3] Grotewald (1959), 347.
[4] Baker (1925). The summary here is from Jonasson (1925).

and consequently commands a price which gives it an advantage over other crops or products.

2. The crop or other agricultural product which has small bulk or weight per unit of value can best bear the cost of transportation and will be grown, consequently, in those regions offering the most favourable physical conditions; but not to the exclusion of more bulky crops, because a certain quantity of these bulky crops that cannot bear the cost of transportation must be grown locally to meet the local demand.

3. The varying seasonal requirements of the several crops and agricultural products for labour tend to diversify the agriculture of a region.

4. Not only is it desirable to grow such a combination of crops as equalises the seasonal requirements of labour, but also it is important to grow such a combination of crops as will maintain soil fertility and promote freedom from insects and disease.

5. Opposed to these tendencies toward diversification is a tendency to grow the most productive crops (value per acre) on the most valuable land, and as population increases and land becomes relatively scarcer this tendency becomes stronger, since the more expensive land and more abundant labour must be profitably employed.

6. Lastly, but not least important, is the character of the farm population and the accumulated community skill and experience.

Between 1926 and 1934 *Economic Geography* published twelve articles by O. E. Baker in which he analysed the agricultural regions of North America, while other contributors dealt later with other continents. Baker did not, so far as we know, begin by looking for regions which theory told him should exist, but rather by accepting such as were already widely recognised and by relying on practical knowledge and statistics to reveal others. Baker's concern was to locate and to describe, not to set the regions in their economic relationships, internal or external. Thus, 'Baker's descriptive treatment of the Corn Belt as a going concern stands unrivalled . . . but should, it would seem, better be regarded as a contribution to regional than to economic geography'.[1] The Corn Belt, also termed the Corn and Soy Bean Belt to take account of changing cultivation patterns,[2] is a good example for the illustration of problems of delimitation.[3]

In some cases Baker was able to use boundaries beyond which the appropriate crops were not cultivated at all, for example the northern boundary of his cotton belt. Corn, however, is the most widely grown crop on the continent and what was required was a valid method of accurately distinguishing the area of corn-growing par excellence, *the* Corn Belt. He defined it as including 'that portion of the east central United States in which corn (maize) is produced in great quantities and is more important than any other crop'. This definition would, however, include much of his Corn and Winter Wheat Belt. It has been suggested that what he really meant was the dominant crop, in the sense that corn provides the focus for the whole farm organisation including the crop rotation, but he does not say this.

On the quantitative measure of relative importance of corn that could be mapped he says 'The yearly average production along the southern margin is 3000 bushels per square mile and this is generally also true along the northern and eastern margins.' We are not, however, told by what method he arrived at the figure of 3000 bushels, nor whether he applied it strictly.

A rigorous procedure which is appropriate in a case like this is to make a farm-by-farm investigation along a series of transects from well outside the approximate boundary to well inside it. Correspondences noted between any line joining the points at which corn ceased to be the dominant crop and other boundaries, physical or economic, would have validity as firm correlations, established *after* the extent of the belt had been fixed. No correlation can have any value at all unless the variables it relates are completely independent. Baker's methods may have been sound but he has left no way of judging them. In spite of this failing his contribution was important in furthering the use of statistics.

Realisation of the value of the work being carried out by Dr Baker led to the initiation of a more detailed study of the farming types of the United States. Data were obtained through the 1930

[1] Buchanan (1959), 10.
[2] Haystead and Fite (1955).
[3] Summary based on Buchanan (1959).

census of agriculture and all farms were classified into twelve types and five subtypes on a basis of farm income. Further processing led eventually to the drawing of boundaries defining 812 type-of-farming districts, which were subsequently grouped into regions and provinces.[1]

Also in this period Hartshorne and Dicken classified the agricultural regions of North America and Europe, using statistics for delimitation.[2] This appears to have been the first attempt to analyse the agriculture of the two continents on a basis of statistical measurement. The limiting criteria were selected from a range of isopleths based on a method suggested by W. D. Jones[3]:

> For any particular limit, that isopleth was selected which seemed to represent most closely the change from one definite type to another. In a number of cases that one was selected which conformed most closely to the boundaries based on value of products . . . the ratio based on acreage was regarded as more significant geographically than one based on value, not only because it is an actual area measurement but also because it fluctuates much less widely from year to year.[4]

The table of criteria used for this classification is given below. It is instructive to study the basis of the delimitation of the regions and also to compare the types of agriculture with those defined by Whittlesey and discussed in Chapter 4.

In its use of statistics to provide a scientific measurement basis for classification in agricultural geography, applicable to diverse regions, this article represented another major step forward. On the other hand, it failed to recognise the difficulties inherent in combining crop and livestock data and the importance of distinguishing the object of crop growing, e.g. for sale or consumption as fodder on the farm.[5]

British geographers began using statistical data to assist in the delimitation of types of agriculture, which had hitherto been discussed mainly by reference to physical or 'natural' regions, as early as the First World War period. The first Agricultural Atlas

[1] Elliott (1933), quoted by Renner (1935).   [4] Hartshorne and Dicken (1935).
[2] Hartshorne and Dicken (1935).   [5] Chisholm (1964).
[3] Jones (1930).

TABLE 6

CRITERIA FOR TYPES OF AGRICULTURE

	Major Crop	Others	Limits
All types	...	...	Crop-and-pasture land >10% total area.
I Mediterranean	wheat	barley, vines, fruits	Vine and sub-tropical tree crops >15% cropland.
II Corn-wheat-livestock	corn	wheat, oats, hay	Cotton < ½ corn acreage. Tobacco <20% cropland. Corn and wheat >30%, corn alone at least 20% cropland.
III Small grains-livestock	wheat, rye	oats, barley, potatoes, hay	Tilled crops >hay and pasture. Wheat and rye >10% of crop-and-pasture land.
IV Hay-pasture-livestock	hay	oats, barley, potatoes, silage corn	Hay and pasture >tilled crops. Wheat and rye >10% crop-and-pasture land.
V Extensive commercial grain	wheat	rye, corn, barley, oats	Livestock <20 units per 100 acres of cropland. Large farms; low yields. Cropland >20% total area.
VI Commercial orchard and truck	...	...	Orchards and vegetables >20% cropland.

of England and Wales[1] used parish statistics,[2] which were depicted as dot distribution maps. In the following decades a number of studies were made on this basis and both dot and density shading maps based on parish statistics featured as a standard method in the reports of the Land Utilisation Survey of Britain.[3] In these reports statistical methods supplemented field enquiries in the interpretation of the land use maps prepared in the early 1930s. Unfortunately, the boundaries of parishes do not coincide generally with significant physical features, but rather run from river valley on to surrounding upland or even mountain regions. Hence, although England is but a small country, and parishes numerous, the subdivision is not fine enough for satisfactory microgeographical work. Parish statistics permit the clarification of distributions over broad regions and changes within these areas over time, but for study of agricultural regions in detail in areas where there is great physical and cultural diversity the network of statistical units must present a very fine mesh.

[1] Howell (1925), 2nd ed. Messer (1932).
[2] Collected by the Ministry of Agriculture and Fisheries, also by Department of Agriculture for Scotland.
[3] Stamp (ed.) (1937–44).

Such a fine mesh existed and was made use of in Northern Ireland. There, the agricultural statistics were formerly collected by the police and returned by enumeration districts of which there were 550 in the six counties of Northern Ireland, an average size of 2450 hectares. Statistics were published[1] only in lesser detail but for years in which this method was used the details by enumeration districts were accessible to research workers. By their aid quite detailed distribution maps could be constructed and the maps in the report of the Land Utilisation Survey of Northern Ireland[2] were compiled on this basis. Even with this degree of detail there were still difficulties, such as enumeration districts, like British parishes, including both lowland and hill areas, but the error in the resulting maps was small compared with those constructed with parish statistics. Unfortunately, this method of collection and processing of statistics is no longer used and similar detail is not available for research purposes. The maps in Figure 14 show the degree of detail obtainable with the enumeration district basis as compared with the present rural district basis.

In surveys like those carried out by the government departments responsible for agriculture in the United Kingdom, statistics of crops, livestock, labour employed and many other details are required under legal duress every year from every farmer. But the law also requires that such statistics shall be regarded as completely confidential so that they may not be disclosed to even a *bona fide* research worker. This is unfortunate, since they would provide excellent material for analysis. Since the farm, not the parish or enumeration district, is the unit of operation and the financial unit, it is obviously a desirable point of departure for research in economic geography as well as in agricultural economics. Many studies based on selected farms and random samples of farms have been carried out by agricultural economists but comparatively few by geographers. Most of the surveys which have used data for a large number of farms have been official,[3] and consequently access to the official

---

[1] Regular publications of the Department (formerly Ministry) of Agriculture for Northern Ireland.
[2] Symons (ed.) (1963).
[3] For example, *National Farm Survey of England and Wales (1941–1943)* Summary report, H.M.S.O., 1946; *Types of Farming in Scotland*, H.M.S.O., 1952.

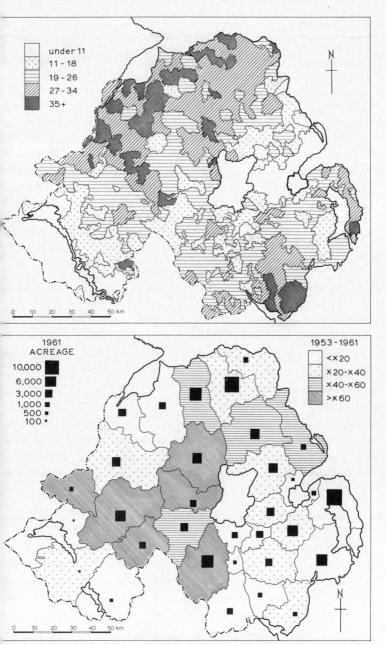

Figure 14. A comparison of detail in density maps obtainable with former and current agricultural census data for Northern Ireland.
(A) Ploughed land as a percentage of crops and pasture, 1953, by special enumeration districts, subsequently discontinued.
(B) Increase in cultivation of barley, 1953–61, by rural districts.

Source: L. Symons, *Land use in Northern Ireland.*

statistics has been permissible, though details for individual, identifiable, farms are never published.

Realising the limitations of parish statistics, even when interpreted with the aid of fieldwork, Birch[1] for a survey of farming-type regions in the Isle of Man decided to base this on a sample of farms selected for the purpose. He notes that any such survey must incorporate (i) an objective method of making a representative selection from the total of farms in a given landscape; and (ii) a method of obtaining from those farms a standardised body of quantitative and qualitative data.[2] He used the 1 : 63,360 and 1 : 10,560 maps as 'statistical frames'[3] and 232 farms out of the total population of 1160 Manx farms were selected.

> They were spaced, as far as possible at regular intervals, over the farming landscape. Of this original selection, 14 farms or 6 per cent, proved unco-operative. A further 24 farms were added as the survey proceeded. This was necessary because the original density of selection did not appear adequate to define what were emerging as localised areas of dominant milk production adjoining the smaller centres of population, and equally localised pockets of store cattle raising near the moorland edge.

This illustrates the commonly experienced need to modify the statistical sample as experience is gained in the field in spite of careful reconnaissance survey. Birch continues with a statement of the two qualities in the sample ensured by the systematic selection of farms from areal frames:

> (i) An adequate spatial distribution of the farms of all sizes. This is essential.

> (ii) The greater proportional representation of the large holdings. This accords with the design of the sample used in the National Farm Survey. It was made possible in this case because the probability of a selection of a farm, by this method, is proportional to its size.

Each farm selected for the sample was visited and if the farmer agreed to co-operate, a questionnaire was left for completion.

---

[1] Birch (1954).   [2] Birch (1954), 144.   [3] Birch (1954), 145.

'110 questionnaires, for 45 per cent of the total farms, were completed in this way, though in some cases after as many as three visits to each farm, extending over a period of two years. For a further 55 farms, 23 per cent of the total questionnaires were completed in full by students or by the writer'.[1] Some data were collected for most farms for which the full questionnaire was not completed.

This report indicates the difficulties experienced by the private research worker, with limited time and resources, in obtaining data from a sample of farms by individual interview and questionnaires. It goes a long way towards explaining why such investigations are not common.

This problem was encountered in a marked form in the compilation of the report of the Land Utilisation Survey of Northern Ireland. It was felt that the interpretation of the land use map made in 1938–39 and further analysis of the land use pattern demanded information on a large number of individual farms. Although Northern Ireland is a small area, only about 13,500 square km (8 million hectares), the number of farms amounted in 1957 to about 74,000. Workers engaged in the different areas were given freedom to use their own methods to achieve this object. Postal questionnaires were not expected to be of great value, but the research workers concerned with some counties tried using them. 441 questionnaires were sent out to two counties, but of these only 87 were returned completed. Reliance on personal field enquiries was therefore continued.

Advances in Statistical Analysis

While the collection of data on the farm-unit scale continues to lag, national and international statistics on crop acreages, numbers of livestock and other details of agricultural land occupance grow rapidly in bulk. It is therefore fortunate that the digital computer makes possible the handling of an amount of data that would have been impracticable without its aid. This permits the combination of data relating to different variables to resolve the patterns of complex distributions. An advance in this type of work was made by Weaver in his assembly of data for several different crops produced in association in the Middle West of the United States.[2]

[1] Birch (1954), 146–147.   [2] Weaver (1954).

Basing his work on acreage statistics, Weaver computed the percentage of total harvested cropland occupied by each crop that held as much as 1 per cent of this total in each of the 1081 counties covered in his work. Ranked by size within each county, many critical crop combinations were obvious, e.g. corn (C) 33 per cent, hay (H) 33 per cent, oats (O), 30 per cent, mixed small grains 1 per cent in Houston County, Minnesota. Other counties, however, as would be expected, showed all gradations between distinctive crop combinations. It was necessary to devise 'a rigorous approach that would provide objective, constant and precisely repeatable procedures and would yield comparable results for different years and localities'.

To provide a standard measurement a theoretical curve was employed as follows:

monoculture	= 100% of total harvested cropland in one crop
2-crop combination	= 50% in each of two crops
3-crop combination	= 33·33% in each of three crops and systematically through to
9-crop combination	= 11·11% in each of nine crops
10-crop combination	= 10% in each of ten crops.

To measure the actual occurrence of percentages against the theoretical curve, the standard deviation was used

$$SD = \sqrt{\frac{\Sigma d^2}{n}}$$

where d is the difference between the actual crop percentage in a given county and the appropriate percentage in the theoretical curve and n is the number of crops in a given combination.

In fact, the relative, not absolute, values being significant, square roots were not extracted so the actual formula used was

$$\sigma^2 = \frac{\Sigma d^2}{n}$$

To demonstrate the technique, the example given by Weaver is reproduced below. The actual distribution of crop percentages in Keokuk County, Iowa, in 1949 was C 54, 0 24, H 13, S (Soybeans) 5, W (wheat) 2. This pattern measured against the theoretical base curve yields the results given in Table 7.

Table 7

STANDARD DEVIATION ANALYSIS FOR KEOKUK COUNTY, IOWA

	MONO-CULTURE	2 CROPS		3 CROPS			4 CROPS				5 CROPS				
	C	C	O	C	O	H	C	O	H	S	C	O	H	S	W
% of cropland occupied	54	54	24	54	24	13	54	24	13	5	54	24	13	5	2
%, theoretical, base curve	100	50	50	33⅓	33⅓	33⅓	25	25	25	25	20	20	20	20	20
Difference	46	4	26	20⅔	9⅓	20⅓	29	1	12	20	34	4	7	15	18
Difference squared	2116	16	676	427	87	413	841	1	144	400	1156	16	49	225	324
Sum of squared differences	2116	692		927			1386				1770				
Sum divided by number of crops	2116	346		309			347				354				

The deviation of the actual percentages from the theoretical curve is seen to be lowest for a three-crop combination. This result established the identity and the number of crops in the basic combination for the county as COH.

The figures thus derived for the years 1949 and 1939 were plotted on maps for the two years. Boundaries were then drawn around blocks of counties with the same number and identity of crops. The percentage rank of crops within any given group was not considered, e.g. COH, OHC, HOC being consolidated into a common crop-combination region. The problem of small-area speciality crops was dealt with by adding a symbol, e.g. Pc for popcorn, for any speciality crop attaining or exceeding 3 per cent of the total harvested cropland.

The results reveal significant diversity within the long-accepted agricultural 'belts' of the area, to which reference has been made in connection with the work of O. E. Baker:

> . . . So far as the identity of the major crops and the relative amounts of cropland devoted to their use are concerned, the practices in 1949 of the so-called 'Dairy Belt' farmers in southeastern Wisconsin were more nearly akin to those of the 'Corn Belt' hog and beef-cattle feeders of southeastern Iowa than to those of dairy farmers in southern Michigan. In turn, the crop-association use in that year of cultivated land in an average county in southern Michigan found its most clearly defined counterparts in such widely separated 'Corn Belt' lands as northeastern Indiana, northwestern Missouri, and southeastern Nebraska.[1]

These are useful findings which are not made less desirable to have by the fact that the crop associations, like other aspects of the economic scene, are not static but always changing. But, as Weaver himself reminds us, the regions presented pertain only to crops, and only to the land-use associations of crops. They are not substitute agricultural regions.

Weaver's crop combination methods were adapted by Scott[2] to a survey of both crop and livestock combinations in Tasmania. Modifications were made to make the procedure even more 'objective, constant, and precisely repeatable', one being to include speciality crops in the statistical definition, and Scott's aim

---

[1] Weaver (1954), 189–190.   [2] Scott (1957).

was to employ the results to help define agricultural regions. He noted that:

> . . . a study of the crop and livestock patterns in Tasmania reveals that both the grouped combinations and the ranked combinations are relevant, since it is the ranked combinations rather than the grouped combinations which define the major crop regions and the grouped rather than the ranked combinations which define the livestock regions. This stems from the fact that crop associations are by no means so strong in Tasmania as livestock associations.

A further important step in combining dissimilar data for the indentification of regions has been taken by Coppock, also using a modified version of Weaver's method to produce not only crop and livestock combinations, but also combinations of agricultural enterprises in England and Wales.[1] Coppock did not attempt to use parishes as the basis of his work, as there are over 10,000 of them in England and Wales and even when using a computer there are limitations to the amount of data that can be handled, because extraction from records and punching cards or tape is time consuming. He used National Agricultural Advisory Districts, of which there were 350, each comprising 30 or 40 parishes and several hundred farms. As already noted, the parish itself is regarded as too large a unit for detailed regional work, but some compensation in using still larger areal units was claimed by Coppock in that the advisory districts were fairly homogeneous in size. Nearly 90 per cent fell in the range 31 to 730 square km, and they often contained at least no greater variety of physical conditions than the parishes.

To arrive at the various combinations of crops and livestock, Coppock used the adaptation of Weaver's method developed by D. Thomas.[2] Thus the data for the full range of crops under examination were fitted to ideal values and the squares of the differences then summed. Weaver mapped crop combinations without regard to rank. Regions in the Middle West could be distinguished by the different crops, but in England and Wales it is the different ranking of a few crops that reveals the differences, e.g. on southern chalkland, barley, wheat and oats, in that order; on east Midland clays, wheat, oats and barley. Full account of rank

[1] Coppock (1964a).   [2] Thomas (1963).

of all crops examined would have been too complex but Coppock did take into account rank in recognising the leading crop. Minor crops had to be disregarded in map construction but the resulting map of combinations, reproduced in Figure 15 and a corresponding map of livestock combinations provide a valuable basis for

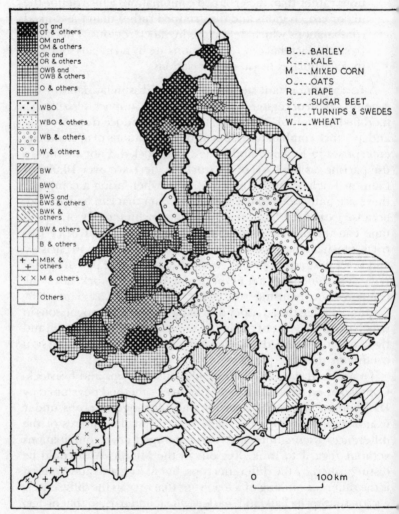

Figure 15. Crop Combinations in England and Wales.
Source: J. T. Coppock, *Economic Geography,* Vol. 40, 1964.

identifying regional types of agriculture. Greater detail of the combinations appears in the *Agricultural Atlas of England and Wales.*[1]

Coppock aimed, however, at not merely plotting the combinations of crops and livestock separately, but the grouping together of these to reveal distribution of types of farm enterprise, which in British farming usually include both crops and livestock on one farm. This involved comparison of unlike units — livestock with crops, and, indeed, different crops, the relative importance of which is not fully reflected by acreage e.g. potatoes and grain. The equating of different classes of livestock, as required for the map of livestock combinations, is, however, fairly simple. Feed requirements provide the normal basis for this type of comparison, and represent a modern equivalent of the souming arrangements employed for centuries by mountain townships with common grazings.[2] Livestock units with slightly varying equivalents are widely used today in calculations of fodder requirements and farming intensity. The factors used by Coppock were: horses, cows, bulls and other cattle two years old and over, 1 unit; other cattle between one and two years old, 2/3; other cattle under one year, 1/3; breeding ewes, 1/5; rams, 1/10; other sheep, 1/15; sows, 1/2; boars, 1/4; other pigs, 1/7; poultry six months old and over, 1/50; poultry under six months old, 1/200.

In order to combine different enterprises other conversion factors are necessary, enabling both crops and livestock to be equated. Monetary values are an obvious possibility but depend on adequate financial information being available. Coppock considers two alternatives: (*a*) standard outputs, in which monetary values are ascribed to each crop and class of livestock; and (*b*) standard labour requirements, in which the annual man-days necessary for each crop or class of livestock provide the common measure. Both measures have to make assumptions based on standards which cannot in fact apply to all farms or all districts. Standard labour requirements were used as the yardstick to measure eligibility for participation in the scheme to assist small farmers.[3] They have many uses, for example in comparing output in a given region with the size of its agricultural labour

[1] Coppock (1964b).
[2] Scottish examples appear in Darling (ed.) (1955).
[3] Cmd. 553, H.M.S.O. London (1958).

Table 8

STANDARD AGRICULTURAL LABOUR REQUIREMENTS

Man-days per hectare

	(a)	(b)
Wheat, barley	8·5	5·0
Oats, mixed corn	11·0	7·0
Rye	8·5	6·0
Maize	—	7·0
Pulses for stock	10·0	7·0
Potatoes	50·0	37·0
Sugar beet	42·0	25·0
Turnips, swedes	30·0	22·0
Mangolds	52·0	27·0
Other crops	17·0	7·0
Vegetables, brassicas	50·0	50·0
Vegetables, roots	52·0	50·0–75·0
Vegetables, pulses for market for processing	} 31·0 {	75·0–200·0 / 7·0–10·0
Other vegetables	100·0	75·0–125·0
Hops	250·0	170·0
Small fruit	110·0	100·0–200·0
Orchards	62·0	50·0–75·0
Flowers, nursery stock	125·0	125·0–450·0
Glass	3250·0	3200·0
Bare fallow	1·2	1·2
Grass for mowing	5·0 }	1·2–1·8
Grass for grazing	0·6 }	

Man-days per head

	(a)	(b)
Dairy cows	15·0	10·0
Beef cows	4·5	3·0
Bulls	7·0	6·0
Other cattle	3·0	2·5
Sows and boars	4·0	4·0
Other pigs	1·2	1·0
Upland sheep one year old and over	0·5 }	0·2–0·7
Lowland sheep one year old and over	1·0 }	
Other sheep	0·25	—
Poultry 6 months old and over	0·3	0·1–0·2
Poultry under 6 months old	0·1	0·05

force.[1] Coppock decided to use these factors, with some modifications, as reproduced in Table 8, column (*a*). Column (*b*) gives, for comparison, the figures used by the Ministry of Agriculture and Fisheries in 1974.[2]

Once man-days had been calculated for individual crops and classes of livestock it was necessary to allocate them to appropriate

[1] For example, in Northern Ireland, Symons (ed.) (1963), 56–57.
[2] There are changes in classification and it has been necessary to omit much detail in the Ministry figures but the comparability is close enough to indicate the greater efficiency and productivity regarded as normal in 1974 compared with 15–20 years earlier.

agricultural enterprises before enterprise combinations could be identified. Seven enterprises were recognised, viz. dairy cattle, beef cattle, sheep, cash crops, fruit, vegetables, and pigs and poultry, the last two being treated together because they depend largely on purchased feeding stuffs. Lack of detail in the census made it necessary to make certain decisions on apportionment of

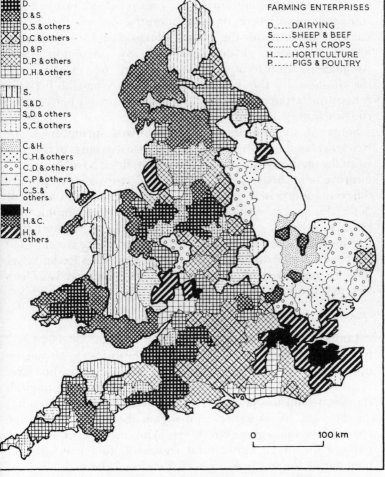

Figure 16. Enterprise Combinations in England and Wales.
Source: J. T. Coppock, *Economic Geography*, Vol. 40, 1964.

cattle to the beef and dairy enterprises, and as to what should be classed as cash crops — wheat, barley, sugar beet and potatoes being so considered. Man-days for fodder (other crops) and grass were then allocated to the different classes of livestock enterprise according to the nutritional requirements of these classes of stock as expressed in livestock units.

A series of maps shows the distribution of leading enterprises assessed in several ways, and combinations of enterprises. In this last (Figure 16) the enterprises are shown in rank order, but in combinations with three or more enterprises only the first two are named. Districts with the same leading enterprise are distinguished by similar shading and a heavy surrounding line. Anomalous features appear in the map; for example, the classification of the Lake District as dairying-with-livestock reflects the layout of large districts, each comprising both upland rough grazings and lowland plain.

Some of the limitations of these maps spring from the coarseness of the mesh provided by the districts, but others derive from the methods used. Coppock suggests three possible lines for improvement; firstly, in the allocation of crops and livestock to different enterprises; secondly, in improving the techniques for determining combinations and, thirdly, in cartographic representation of the combinations. Coppock claims no more than that his maps should be regarded as reconnaissance sketches, but his paper demonstrates the great scope that exists for improvement in the analysis of agricultural distributions and takes a firm step in a promising direction.

Types of Farming and Agricultural Regions

Coppock also points out that his enterprise map is not a type-of-farming map, as it is based upon the district totals and not upon individual farm records. If it were based on individual farm records and boundary lines were drawn on it to distinguish regions, would this be a map of agricultural regions? To put it another way, is an agricultural region the same thing as the type-of-farming region? Whittlesey, in his classic paper[1] set out to indicate the major agricultural *regions* of the earth, drawing boundaries for this purpose which enclosed the major *types* of agriculture, as summarised in Chapter 4. Whittlesey, however,

[1] Whittlesey (1936).

considered that the functioning forms of a type of agriculture included not only the crop and livestock association, but also methods, intensity, disposal of products (subsistence or commercial) and the ensemble of structures used to house and facilitate the farming operations. It thus includes more criteria than would be included in analyses similar to that carried out by Coppock into crop and livestock associations (but applied to individual farms), which he would appear to require to discern type-of-farming regions.

The types of region are different, involving in one case more and wider criteria than in the other. But use of the terms as employed by the respective authors would imply different definitions of 'type of farming' and 'agriculture' which would probably not be generally acceptable.

It is evident that there is need here for attention to terminology with the dual objectives of removing ambiguities and focusing attention on the exact objectives of research projects and their limitations.

Progress in Regional Delimitation and Spatial Analysis

The methods which have been reviewed in this chapter go some way to satisfying the needs of regional division. More advanced statistical methods, however, have been tested extensively by some workers. Examples include the linear programming type of production model,[1] in which limiting assumptions may be relaxed as the models are improved, the use of chi-square analysis for testing regional boundaries[2] and the quantification of similarity as a basis for multifactor regionalisation.[3]

Wolpert used linear programming and regression analysis in the investigation of regional variation in decision making on a sample of 550 farms in Central Sweden, developing through composite analysis a behavioural theory which would be more descriptively accurate than the unsatisfactory normative concept of 'economic man'.[4]

Investigating the effects of village and field layout and resulting transport costs and other factors on decisions on what crops to plant and on the intensity of cultivation in two areas of India, Blaikie utilised a wide array of statistical measures.[5] A hypothesis

[1] Henderson (1957).
[2] Zobler (1957).
[3] Berry (1958).
[4] Wolpert (1964).
[5] Blaikie (1971).

was developed from von Thünen and Dunn stating that each cultivator, having arrived at areal allocations for various crops according to his survival algorithm (which subsumes crop rotation), will grow those crops whose total transport outlays involved in cultivation increase most rapidly with increasing distance, closest to his inputs other than land. A second part of the hypothesis stated that the intensity of production of any one crop will decrease with increasing distance within any one farm, although not necessarily between farms because of variation in farm resources. Blaikie used a detailed questionnaire on basic farm variables and accounts as well as secondary sources and set up an interlocking design using a variety of multivariate statistics. In addition to using simple regression and principal components analysis, Kolmogorov-Smirnov tests were applied to test whether certain crops with varying mean transport requirements were grown in a significantly different distribution from one another in a series of concentric zones around the village core, and in a significantly different distribution from one that would be expected if the crop was grown equally in all zones. An attempt was made also to bring together farm structure variables and field variables in a series of multiple regressions where each group of field variables was averaged for each farm and then treated as farm variables. Sequential multiple regression and best-set analysis were the two chief types of experiment used. Various other simple regressions were calculated to follow up side issues and to illuminate the main avenues of research. The complexity and sophistication of this programme indicate the scope for application of statistical procedures to location patterns and decision-making analysis with the aid of computers, but also emphasise the amount of work that goes into such a detailed investigation of even a small area.

As mathematical models approach the stage of refinement where they can embrace the complexities of functioning and dynamic economic relationships, they must be utilised increasingly in our spatial analysis, though, as noted above, it is not only a matter of methods but also of the nature as well as the accuracy of the raw material that governs the value of the statistical analysis.

Increased use of financial criteria has aided in the recognition of regional variation in farming intensity. Using the ratio of value of farm production to cropland increase, Gregor[1] showed that in California, farming intensity, per acre of cropland, was highest in the desert region where the oases have been rendered exceptionally productive. He also used value added in farming to rank Californian regions, which confirmed the expected lead of the San Joaquin valley but also emphasised the rise of the desert region.

Spatial analysis of financial data may be useful in revealing regional variation in distributions where this is not expected or not intended, as for example, in national policies. The regional effects of public policies in agriculture have been investigated by Bowler, using regression coefficients for time series of subsidy inputs.[2] He concludes that 'both undesirable and unpredictable inputs of grant aid can result from non-selective measures' and, using the evidence of surprising variability of response to calf subsidies and ploughing grants, he argues that more selective direction of inputs by either location or type of farming would ensure an increased efficiency in the use of resources.

To a geographer searching for the most satisfactory way to illustrate and analyse spatial distributions the method used to depict the data is of major interest. Proven but not wholly satisfactory methods such as those used in dot, density and other distribution maps may be supplemented by maps based on more complex mathematical procedures. An example is the use of trend-surface analysis which has been developed for agricultural distributions by Tarrant.[3] The objective is to separate large-scale gradients in a particular variable from smaller, local gradients, resulting in emphasis on the continuity of change that is evident in many aspects of agriculture which may be lost when the data are fitted into discrete regions.[4]

Most techniques used to study problems in agricultural geography are applied mainly in examining relatively limited and local situations, and recognition of such limitations does not imply any criticism of the value of the work. Indeed, it is through the determined tackling of individual problems of agriculture

[1] Gregor (1963).
[2] Bowler (1976).
[3] Tarrant (1969).
[4] For a convenient description of the use of trend-surface analysis as well as consideration of other methods of plotting agricultural distributions see Tarrant (1974).

and agricultural regions by specialists, including geographers employing spatially-orientated approaches, that the most valuable contributions, both theoretically and practically, may be made. Many geographers, however, remain hopeful that the wider issues of understanding and mapping world types of agriculture, which have been shown to present great problems, may eventually yield to persistent efforts in data accumulation, classification and regionalisation, to produce much greater progress than is so far evident since the days of Whittlesey and other pioneers.

Spencer and Stewart contend that agricultural regionalisation, whether of a local territory or the world as a whole, requires the exclusion of any production patterns subordinate to the mode of production selected as dominant for each areal unit. Hence the map cannot effectively depict localised variation or zones of complex mixtures.[1] They favour an approach through classification of systems of agriculture, using relatively few primary genetic components with detailed division by subsets of secondary criteria. They emphasise the importance of cultural factors, believing that economic and technological aspects have been overstressed in past classifications. They set out a tentative hierarchy 'in a loosely evolutionary sequence' as follows:

(1) Shifting cultivation
(2) Sedentary gardening
(3) Primitive plough culture
(4) Formative commercial gardening
(5) Developed general plough culture
(6) Semi-mechanised commercial general farming
(7) Developed mechanised farming

Separate sequences are postulated from (8) traditional pastoralism to (9) commercial livestock ranching and from (10) traditional latifundia to (11) the modern corporate plantation, with further separate categories for (12) industrialised agricultural production and (13) collectivised managerial agriculture. Numerous criteria are specified and tabulated for the identification of systems but the task of allocating actual functioning systems to the categories defined is not attempted, nor is the question of mapping them discussed in this article.

[1] Spencer and Stewart (1973).

The contrasting approach of the International Geographical Union Commission on Agricultural Typology aims at establishing a formal typology with a limited number of objectively quantifiable variables, supported by other significant criteria. Spencer and Stewart consider that it is unlikely that the product of such a practical typology could be translated into a map but Kostrowicki argues that, after a typology has been established, agricultural regions can be delimited easily by generalising the more complicated typological pattern for a given period to a simpler regional picture based on dominance or co-dominance of individual types over a given territory.[1] While it is still too early to judge the full value of the work it would seem that it must contribute significantly to a map of regions of world agriculture even if it does not provide all the material required for a comprehensive map of agricultural systems. Irrespective of the aid to mapping that may eventually be provided, the programme has produced one of the two major examples of international cooperation and sharing of data in the field of agricultural geography,[2] the other being the World Land Use Survey, the contribution of which falls within the next chapter.

[1] Kostrowicki (1976), 244.
[2] Kostrowicki (1974); Reeds (1973).

CHAPTER 10

Land Use and Land Potential

Land use survey formed the spearhead of the advance of geography into the applied sciences[1] as maps of land use became recognised as essential tools of regional planning and development.

British academic geographers created the first systematic and nationally-comparable land use maps in the 1930s.[2] After the Second World War many governments and other national bodies initiated similar surveys for their own countries and maps became increasingly refined and more detailed. Today, there is a complex, high-technology industry catering for international needs in resource survey, largely by air photography and other forms of remote sensing and geographers have many roles in this work.

In a comprehensive study of land use it is proper to include all forms of use, whether agricultural, sylvicultural, industrial or urban, though many studies are concerned with only one or another. We are here concerned only with agricultural land use, which in most occupied regions of the earth is the greatest user in terms of areal extent, though in some regions this is surpassed by the forestal use, and, locally, by urban use. Following the dominance among areal uses of agriculture, many land use surveys have been concerned almost entirely with agricultural use, even though this has not been specified.

As a result of land use survey, areas are grouped into regions in much the same way as has been discussed in relation to agricultural regions, but using different criteria, such as the respective proportions of the different classes of land use,

[1] The majority of land use maps are, strictly, land cover maps, i.e. depicting vegetation, buildings, etc. — the form as against the function. The term 'land use' (or 'land utilisation') is, however, more generally used and may, in practice, be employed for either the land cover or the function to which it is devoted, embracing, for example, both grassland and grazing, arable land and cropping.
[2] Board (1968).

including forests, urban areas, transport installations, industries and unused land. They may coincide closely with agricultural regions, even exactly (if there are no uses of land other than agricultural in the particular region) but the criteria used for definition and delimitation are different — they are land use criteria.

For practical purposes, especially where a study is addressed to a wide variety of readers of varied technical backgrounds and interests, it may be desirable to integrate agricultural and land use regions, to avoid tedious repetition and unnecessary pedantry. This was done deliberately in the report of the Land Utilisation Survey of Northern Ireland. Thus the description of the Lough Neagh Lowlands (given here in Chapter 5) as a region of small, mixed farms, is derived from the wider description of land use regions. Similarly, from a land use region which contained a significant element of forests, one could abstract the information relevant to forestry.

Land use surveys should always be prepared with the needs of a wide range of specialists and laymen in mind, for a land use survey is an expensive undertaking and everyone today is in some way concerned with land use. Although it was the first of the modern land use surveys, that of Great Britain undertaken in the 1930s achieved this quality, and the value of a broad interpretation was proved over and over again in the uses found for it in planning wartime agricultural development and post-war reconstruction.

The Evolution of Land Use Survey

Concern with the uses to which land is put must be as old as agriculture, and the systematisation of knowledge relating to land use dates back at least as far as the application of taxes to land according to its use. One of the best ancient examples was the Norman Domesday Survey of England. Centuries later the agrarian revolution led to surveys which examined usually both agricultural techniques and general aspects of land use county by county,[1] and settlement in the 'new' countries stimulated interest

[1] The first *Statistical Account of Scotland*, which provided systematic studies of this kind, was begun in 1791, similar *Statistical Surveys* of the Irish counties appeared in the early years of the nineteenth century, and many volumes appeared on the English counties, giving in all a most valuable record of the eighteenth and nineteenth centuries in the British Isles.

in assessment of land resources. It was not, however, until 1919
that the idea of a map portraying the use of land systematically
was put forward[1] and not until another decade had passed was
there an attempt to initiate such mapping on a national basis.
Steps to formulate a scheme applicable to the whole of Great
Britain had been made successfully by 1930 and the fieldwork
began in 1931.[2]

In this pioneer scheme costs had to be kept to the minimum and
it was essential to use a scheme that was simple and readily
understood by voluntary workers with varied or with little
training. The scheme evolved, nevertheless, made possible the
presentation of a vivid picture of the contrasts in land cover
throughout the country. The field mapping was done on
Ordnance Survey maps on the scale of six inches to one mile
(1:10,560) which were subsequently reduced to one inch to one
mile (1:63,360). Most one-inch sheets were printed and published, the exceptions being some of the sheets of the Highlands
of Scotland where almost all of the land was used as rough grazing
and deer forests.

The classification used, with the letter placed on the field sheets
and the colouring used on them and on the published maps, was
as follows:

(1) Forest and woodland .... F Dark green
(2) Meadowland and permanent grass ... M Light green
(3) Arable or tilled land, fallow, ... A Brown
    rotation grass, and market gardens
    [market gardens, where clearly so, ... A (M.G.)]
(4) Heathland, moorland, commons, ... H Yellow
    and rough hill pasture
(5) Gardens, allotments, orchards, ... G Purple
    nurseries, etc.
(6) Land agriculturally unproductive, ... W Red
    e.g. buildings, yards, mines,
    cemeteries, etc.
(7) Ponds, lakes, reservoirs, ditches, ... P Blue
    dykes, streams and anything con-
    taining water

Forests were subdivided into high forest, coppice, scrub, any

[1] Sauer (1919), 48.   [2] Stamp (1948, 1962).

cut down and not replanted, and whether coniferous, deciduous or mixed.

The classification of rotation grassland as arable was correct in terms of land management but gave rise to some difficulties in interpretation. In the subsequent survey of Northern Ireland in 1938–39 all grassland, whether in rotation or in permanent pasture of good quality, was grouped together. A category was also introduced for lowland bog, which was not a significant feature in Britain but of considerable extent in Ireland.

Reports were written describing the land use of each county in Britain and published in 92 parts between 1936 and 1946. The time between the beginning of the field work and the completion of the reports may appear excessive, but to appreciate the difficulties before and during the war which had to be overcome it is necessary to read the account of the history which introduces the summary volume, *The Land of Britain, its Use and Misuse.*

It is instructive to compare the later land use survey of England and Wales with the old. A sample area is compared, with the addition of soil and relief maps, in Figure 17. The newer land use map is much more detailed, which is made practicable by publication at the scale of 1/25,000. Thirteen main groups are used, with a crayon number specified for each to avoid confusion[1]:

(1)	Settlement (residential and commercial)	Grey
(2)	Industry	Red
(3)	Transport	Orange
(4)	Derelict land	Black stipple
(5)	Open spaces	Lime green
(6)	Grass	Light green
(7)	Arable	Light brown
(8)	Market gardening	Purple
(9)	Orchards	Purple stripes
(10)	Woodland	Dark green
(11)	Heath and rough land	Yellow
(12)	Water and marsh	Light blue
(13)	Unvegetated land	White

All classes except (3), (4), (5) and (13) are subdivided. The

[1] Coleman and Maggs (1961).

Figure 17. A comparison of British land use, soils and relief maps of an area in the Chiltern Hills between Wendover and Great Missenden to show the growing detail recorded in contemporary land use mapping and the relationships between land use and physical conditions. (Scale of these reproductions is approximately 1 : 30,000).

A Part of Sheet 106 of the Land Utilisation Survey, surveyed 1931–32: A, Woodland; B, Arable; C, Permanent grass; D, Heath and rough pasture; E, Gardens and orchards.

B Sheet 264 of the new survey of 1960: *a–e*, Arable (*a* cereals. *b* lev legumes. *c* roots. *d* green fodder. *e* fallow. *f–h*, Market gardening (*f* field

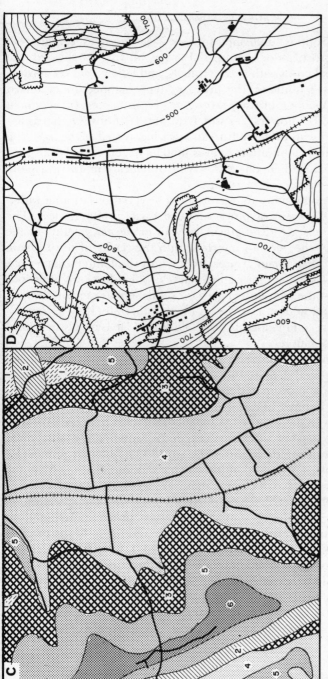

C Soils Sheet 238, 1961: 1. Rendzina (parent material, Chalk); 2. same, steepland phase; 3. Brown Calcareous Soil (on flinty and chalk Head over Chalk); 4. and 5. Brown Earths (4. on loamy and gravelly Head over Chalk, 5. Clay-with-flints over Chalk; 6. Gleyed Brown Earth (on Plateau Drift). Crown Copyright reserved.

D Topography, abstracted from the Ordnance Survey Map, New Popular Edition Sheet 159, 1945 (full revision, 1930). Crown Copyright reserved.

arable group is subdivided into six — ley legumes, cereals, root crops, green fodder, industrial crops, all further divided into individual crops with letter symbols, and fallow. Market gardening is divided into (*a*) the 'ordinary' type, mapped under 'mixed crops', potatoes, brassica crops, etc., (*b*) nurseries, (*c*) allotment gardens, (*d*) flowers, (*e*) soft fruit and (*f*) hops. Orchards are recorded with underculture and grazing as well as the tree crop itself. The complications of grassland management led to all grass being included in one class, with modifications for grassland infested with scrub or rushes.

The subdivisions are represented by variations of tone within the main colour, or other subdued cartographic devices. The subdivisions are distinguishable with moderate scrutiny of the maps but they do not interfere with the clarity of the 13 main groups. These main groups are not only similar in most respects to the classes of the original survey, but differ only in minor details — the adoption of grey for settlement as used on the base maps, and addition of classes (3), (4), (5) and (13) — from the scheme recommended for the Old World Division of the World Land Use Survey.

The World Land Use Survey Commission of the International Geographical Union functioned from 1949 to 1976, during which time it promoted the making of maps and reports on land use in a number of countries and aimed to produce a world land use map on the scale of 1 : 1,000,000. Such a map would provide a valuable step towards a much needed inventory and evaluation of land resources throughout the world. The limited number of basic classes proposed in the scheme was intended to obtain the necessary degree of comparability in surveys, with subdivision as required for detailed local purposes. Even the basic classification of nine categories presented considerable problems in simplification of surveys for reproduction. The reduction of the original land use maps of Britain to a scale of 1 : 625,000 showed the degree of generalisation required. At this scale the smallest area that could be shown separately was about 40 hectares, so adjacent blocks of fields had to be run together, with the proportions being preserved.[1] On a scale of 1 : 1,000,000 generalisation would have to be undertaken to produce minimum unit areas of at least 100 hectares.

[1] Stamp (1948, 1962), 33.

Maps on the 1:1,000,000 scale have now been prepared for a number of areas but the W.L.U.S. scheme has been largely overtaken by events. The rapid development of remote sensing from aerial and satellite platforms and the growth of land survey work into a major industry offering substantial rewards to highly capitalised commercial companies, operating on an international basis, have facilitated the rapid production of maps for individual national requirements. By no means all these maps are published and lack of standardisation arises from the multitude of authorities involved, as well as differing requirements and techniques and there has been little attention to the classification proposed by the World Land Use Survey, but at least a large part of the world can now be said to be mapped on small scales with selected development areas studied in, often, quite detailed surveys.

Although the mapping of large areas is now most effectively and most economically carried out by technologically advanced remote sensing there is still a place for field-by-field survey on the ground for detailed mapping of crops and multiple uses of land, the impact of city growth and transport networks and similar complex matters.

That land use survey has a value for practical purposes is no longer questioned. Accounts of its application to physical planning are numerous. An interesting volume which made possible comparison of the British survey with the development and aims of land use mapping in a socialist economy was produced from the proceedings of the first Anglo-Polish Geographical Seminar.[1] The essence of the Polish approach is contained in the following extract:

> Research connected with land utilisation has two objects, scientific and practical, but it is difficult to separate them. The most general scientific aim is, above all, the study of the ways in which man's economy utilises its natural environment. This is essentially a geographical study, which can be greatly helped by land utilisation survey . . . the survey can serve as an important foundation for the drawing of conclusions aimed towards a more rational utilisation of the geographical environment.

[1] Polish Academy of Sciences (1961).

The Polish map is designed to show four aspects of the work:

(1) The *form* of land utilisation, the actual use of the land (for which the term land cover may alternatively be used) following the World Land Use categories as far as possible.

(2) The *subject* of land utilisation, i.e. the farm holdings, are shown by property boundaries.

(3) The *way* of land utilisation, such as methods of crop rotation, the use of fertilisers and mechanical aids to cultivation, shown by black symbols.

(4) The *directions* or *orientation* of land utilisation, the objects of production, e.g. fodder crop orientation with preponderance of animals, by colour gradations within the main forms.

Thus it will be seen that the Polish survey takes a broad view of the scope and purpose of a land use survey, and shows not only land cover and land use in the sense that these terms are used here but also background data such as size of farms. Normally, such data are presented in separate maps, often as transparent overlays which can be used together with similar overlays of physical conditions, transport capacity and costs, etc. to facilitate understanding of interrelationships. The presence of such details on the basic land use map indicates the object of the Polish map to assist in practical planning.

A feature of land use maps which must be stressed is that the presentation of material on them is objective, i.e. they record facts with a minimum of interpretation on the part of the surveyor and cartographer. This is so even in a broad survey like the Polish one. But, because land use maps record the land use only at a moment of time, they cannot do more than present a synoptic picture of the landscape. A field which is under grass this year may be growing a cereal crop next year. Hence, in one sense the map is out of date before it is printed. This failing of the land use map has sometimes been offered as a reason for not undertaking such mapping, but as a criticism of technique it fails to take account of the stability of the pattern in a larger area, such as the farm unit, or a typical square mile of territory. The change in the individual field is often merely an expression of crop rotation, and is offset by corresponding changes in other fields. The stability of the pattern even in a country of fairly rapid technological change like Great Britain is proved by comparison of maps of different dates.

A land use map may be considered a valid document for a number of years, just how long being dependent on the nature and rate of change. Sample surveys of land use in restricted areas and the interpretation of census statistics are necessary to determine when a new survey is needed. Resurvey may be needed very frequently on the fringes of urban development, but only at long intervals in rural areas.

The more information that is included on a land use map, the sooner it is likely to be out of date. Whereas a pattern of arable and grassland may remain basically unchanged over many years, the crops in the fields will change much more, while the progress or health of the crop will change constantly. This type of information is, however, demanded increasingly, particularly since chemical methods of disease suppression and application of artificial fertilisers have become more common. Air photography is virtually the only way of providing information rapidly enough to be of value for such aspects of production control and the combination of colour and false colour or infra-red photography with black-and-white panchromatic films makes possible wide ranging interpretation, indicating the need for crop spraying or fertilising and, by re-survey (as in experimental situations), the results of such treatment.[1]

The feature of objective recording means that there is on a land use map no indication of whether or not the land is being used in a way appropriate to its quality, except in so far as this may in extreme cases be obvious from, say, relief relationships. Thus, the colour yellow, depicting heathland and rough grazing on mountains in the Highlands of Scotland, may be taken to indicate satisfactory utilisation at 1000 metres, but in the valleys between the mountains it poses the question of whether here might not be conditions suitable for cultivation. Again, there is nothing on a land use map, normally, to indicate whether an area of grass is very fertile or poor, without being poor enough to be classed as rough grazing. Nor is there any indication of whether the natural conditions are such that substantial improvement of the land is practicable.

Land Potential Maps

It is a matter of general knowledge that some land is better than

[1] Bell (1974).

other land in the same use, and with the pressure of growing population in relation to a more or less fixed supply of land it is important that variations in land quality should be studied and mapped. This is partly the prerogative of the soil surveyor, and a full inventory of land resources can be achieved only with detailed soil survey. Soil survey, however, is often directed to pedological rather than productive aims, and in any case needs re-interpretation to serve the needs of the planner. Soil, too, is not the only physical factor in the productivity of land. To meet the need in assessing land resources many schemes of classifying land potential have been developed.[1]

In classifying agricultural land potential the object is to establish the fertility or potential fertility of each tract of land. The surveyor seeks to answer the questions

(1) For what use is this land most suitable?
(2) What is the potential productivity of this tract of land compared with other tracts?

Classification may be of (a) a general nature or (b) specific. A general classification provides basic information which will assist planning of various kinds, e.g., in questions of land use involving choice between one use and another whenever they may arise. Thus the classification might be used at one time to help decide whether a particular tract being used for low intensity grazing should be used for a reafforestation scheme or for improved agriculture or left undeveloped, and at another time as a guide to whether the same tract should be reserved in its existing use, or whether building construction should be permitted. The essence of such a classification is a summary of the inherent physical characteristics not only of the tract under consideration but of other areas which could provide alternative sites for development. Such a classification has the advantage that once it is completed it can be referred to without delay to assist in the solution of many planning problems. Indeed, no country can be considered to be adequately mapped until a land classification map of this kind on a sufficiently large scale is available. Nevertheless, such a classification must be to some extent

[1] U.S.A. National Resources Planning Board (1941), Jacks (1946), Ministry of Agriculture, Fisheries and Food (1974), Stewart (ed.) 1968.

subjective and imprecise because of the many factors involved, and other classifications are needed.

A group (*b*) classification is prepared with a specific object in mind, e.g. to classify land into grades of suitability for afforestation, or for agricultural development, or even for a single species of tree or crop. Such classifications are easier to make than a general classification and may be made first but a general classification should be regarded as a further objective. Thus, given a map showing suitability of land for cattle grazing, it can be seen whether a given tract is classified as suitable or unsuitable for that purpose. If, then, there is available a general classification map this will show whether, compared with other areas, its general fertility is rated high or low. Although not suited to cattle grazing the tract may be rated as first class land because of suitability for arable crops. This type of single-subject potential map will not be discussed separately here. The problems it poses are similar to those of the general map in restricted form. It may, however, be noted that many classifications do not fall with complete exclusiveness into either category.

The land capability classification devised by the United States Soil Conservation Service is an example of a classification which is concerned mainly with the erosion hazard, but which introduces other aspects of land potential and which has been used for national schemes in other countries so that it might form the basis of international assessment of land capability. In the original scheme, eight classes are recognised, with a primary division into land suited and land not suited to cultivation (four classes each) and suffixes are added to indicate particular limitations such as erosion risk or wetness.[1] The main classes are:

Class I. Very good land that can be cultivated safely with ordinary farming methods.

Class II. Land that may be cultivated safely with moderate precautions. Soils may lack depth, be liable to wetness or present other problems but not to a serious extent.

Class III. Land which has considerable limitations of use, but which may be cultivated regularly if hazards are guarded against. Characteristics include moderately steep slopes, high susceptibil-

[1] U.S.A. Department of Agriculture (1951), (1954), summarised in standard conservation texts.

ity to erosion, hardpan or claypan, and very impermeable or sandy soils.

Class IV. Land with very severe limitations in use. Cultivation should be limited to occasional crops with extreme care being exercised. A long rotation of 5 or 6 years in grass followed by a crop of grain or lucerne is often practicable. In some semi-arid regions the best land is in Class IV. Most Class IV land in humid regions is well suited to forestry.

Class V. Land which although nearly level is not suitable for cultivation because of wetness, stoniness or other factors. Forestry and grazing are suitable uses with few limitations.

Class VI. Land which is steep, rough, dry, wet or otherwise unsuitable for cultivation. Some Class VI land can be tilled just sufficiently to establish pastures and some can be used safely for tree crops. Grazing must be restricted to safe numbers of stock and to appropriate periods. Gully control, contour furrows, ridges, water diversions and other conservation measures may be needed.

Class VII. Severe limitations or severe erosion hazards under grazing or forestry uses characterise this land. Depletion of cover leads to more rapid erosion than on Class VI land, and structures such as contour furrows and ridges cannot be used because of steep slopes, shallow soils or other unfavourable factors. Forestry is preferred to grazing for conservation purposes.

Class VIII. Forestry as well as grazing and cultivation are regarded as unsuitable on this class of land. It includes marshes, deserts, badlands, and high mountain land, and is suited only for wild life preservation, recreation and water catchment, with appropriate watershed protection measures.

This classification has evident value for land where soil erosion or other hazards limit cultivation, but requires substantial modification for the assessment of productivity in any but the most general terms. To aid in the comprehensive planning of land use it is desirable to incorporate allowances for all the significant factors of the environment. A land classification scheme normally seeks to express the inherent, natural quality of the land. To the extent, however, that the effects of past cultivation have modified the natural fertility of the soil they may be taken as having become absorbed into the inherent quality of the land. The comparative quality of land may be assumed to be assessed at its normal, existing level, which could only be upgraded, if at all, by abnormal investment, such as major reclamation schemes.

Land Classification in Britain

In Great Britain, the Land Utilisation Survey played the major role in the production of the first national map of land potential, completed in answer to urgent wartime needs. The subject was discussed at a series of meetings held in 1943 by specialists in several sciences with the aim of producing urgently a simple classification of land which could be used as a basis for national planning, especially for the delimitation of the tracts of good agricultural land to be avoided, if possible, in building developments.[1] The Soil Survey's definitions of good, medium and poor quality land — Major Categories I, II and III — were accepted by all, and these were subdivided into ten classes already adopted by the Land Utilisation Survey. This system facilitated the rapid compilation of a national map on a scale of 1:625,000 from the land use maps and other available data with a minimum of fieldwork.

After the war, when the urgency was less, regional surveys were undertaken on an *ad hoc* basis to provide land classification maps for regional planning.[2] Most of these local schemes used the three major categories but in Northern Ireland and Scotland a preference for four major categories was arrived at independently because of the small extent of first class land (Category A) and the large areas of medium land (B), poor — just cultivable or marginal — land (C) and very poor mountain and other land suitable only for rough grazing (D).[3]

During the 1960s the Agricultural Land Service of the Ministry of Agriculture, Fisheries and Food and the Soil Survey developed related but essentially independent classifications for agricultural land in England and Wales in the former case and for Great Britain as a whole in the latter case. The Agricultural Land

[1] Stamp (1948, 1962), 353.
[2] For example, by the West Midland Planning Group in *Conurbation*, the University of Bristol in *Gloucester and Somerset, A land classification*, and by the North East Development Association in *A physical land classification of Northumberland, Durham and part of . . . Yorkshire.*
[3] Symons (ed.) (1963).

Service was concerned especially with the provision of maps to help to guide planning decisions so that a minimum of good agricultural land should be built over or otherwise removed from agricultural use. It thus sought a successor to the wartime map and as a scale large enough to show reasonable detail and to allow incorporation of amendments was required the 1:63,360 scale was chosen, but speed of production dictated only a reconnaissance field survey and the early maps were marked 'provisional'.[1] Meanwhile the Soil Survey began more detailed maps, classifying land in terms of use capability parallel with the mapping of soils and incorporating climatic features.[2] The resulting maps are also published on the 1:63,360 scale and are labelled 'provisional'.

Both classifications are defined in terms of the potentialities of the land for agriculture and the limitations imposed by soil, relief, climatic and other physical factors in conformity with the U.S. Land Capability Classification.

The Agricultural Development Advisory Service (Lands), successor to A.L.S. uses only five grades, of which the first four are defined in terms very similar to those used for Classes 1 to 4 of the Soil Survey's 7-class scheme. As the latter is the more comprehensive, brief descriptions of its classes are given below, followed by comparison with the A.D.A.S. and American classifications.

Class 1. *Land with very minor or no physical limitations to use.* Soils are well drained deep loams, related humic variants or peat with good moisture conditions, well supplied with plant nutrients or responsive to fertilisers. Sites are level or gently sloping and climate favourable. A wide range of crops can be grown and yields are good with moderate inputs of fertiliser.

Class 2. *Land with minor limitations that reduce the choice of crops and interfere with cultivations.* Limitations may include, singly or in combination, slightly disadvantageous drainage, soil depth, structure, or texture, slope, slight vulnerability to erosion or slightly unfavourable climate. A wide range of crops can be grown though some difficulty may be found with harvesting root crops and winter-harvested crops.

Class 3. *Land with moderate limitations that restrict the choice o*

[1] Ministry of Agriculture, Fisheries and Food (1966), Morgan in MAFF (1974).
[2] Bibby and Mackney (1969), Mackney in MAFF (1974).

crops and/or demand careful management. Owing to limitations resulting from any of the adverse factors the range of crops on this land is generally restricted to grass, cereal and forage crops and the timing of cultivations may be affected.

Class 4. *Land with moderately severe limitations that restrict the choice of crops and/or require very careful management.* Limitations include (1) poor drainage difficult to remedy, (2) occasional damaging floods, (3) shallow and/or very stony soils, (4) moderately steep gradients, (5) slight erosion, (6) moderately severe climate. Climatic disadvantages combine with other limitations to restrict crops to, mainly, grass, with oats, cereals and forage crops as possible alternatives.

Class 5. *Land with severe limitations that restrict its use to pasture, forestry and recreation.* Limitations are due to defects in drainage, flooding, slopes, severe risk of erosion, severe climate, etc. which cannot be corrected. Arable cropping is impracticable but mechanised pasture improvements are feasible. The land has a wide range of capability for forestry and recreation.

Class 6. *Land with very severe limitations that restrict use to rough grazing, forestry and recreation.* Limitations are sufficiently severe to prevent the use of machinery for pasture improvement. Very steep ground which has some sustained grazing value is included.

Class 7. *Land with extremely severe limitations that cannot be rectified.* Such land includes (1) boggy soils, (2) boulder-strewn soils, bare rock, scree or beach sand, (3) untreated waste tips, (4) very steep gradients, (5) severe erosion, (6) extremely severe climate. Exposure, protracted snow cover and a short growing season preclude forestry though some rough grazing may be available seasonally.

The A.D.A.S. Grades I to III are defined in almost identical terms to the above Classes 1 to 3 and there is close similarity of the A.D.A.S. Grade IV and the Soil Survey's Class 4. Of these, the main agricultural lands, the A.D.A.S. further notes that land above 400 feet (about 120 m) which has more than 40 inches (say 1000 mm) annual rainfall (45 inches in western regions) or land with a high proportion of moderately steep slopes (1 in 8 to 1 in 5) will generally not be rated above Grade III, which accords closely with the Soil Survey's guidelines,[1] while land above 600 feet

[1] Bibby and Mackney (1969), 10.

(about 180 m) which has over 50 inches (say 1250 mm) annual rainfall or has a high proportion of steep slopes (between 1 in 5 and 1 in 3) will generally not be graded above IV.

The A.D.A.S. Grade V is described as applying to land with very severe limitations, and areas over 1000 feet (300 m) with over 50 inches (say 1250 mm) annual rainfall or land with a high proportion of very steep slopes (above 1 in 3) will not normally be graded higher than Grade V. The Soil Survey's Classes 5–7 allow more differentiation of the poorer lands. There would appear to be no reason why the two schemes should not be amalgamated with benefits for most users.

Following U.S.D.A. practice, suffixes are used by the Soil Survey to denote the particular limitations affecting individual areas, and specific information is assembled by land use capability units, within which soils respond similarly to management and improvement practices and have comparable yields.

It will be seen by comparison with the U.S.D.A. scheme given above that the British classifications compare fairly closely with the American usage in classes 1 to 4. The U.S.D.A. Class V, allowing mainly for wet soils in level sites, was omitted as such lands could be dealt with in other classes as appropriate, a modification suggested by Kellogg[1] and adopted in Canada.[2] All eight classes are, however, used in the New Zealand adaptation of the classification.[3]

In spite of the variations which occur in the interpretation of the classes, the U.S.D.A. scheme appears to offer the greatest possibilities for some measure of international use, though it must be remembered that direct comparability is limited. Comparability is at its greatest within broad climatic zones and an attempt to classify other temperate lands — the main producing areas for bread grains and important classes of livestock — in comparable land capability terms[4] should be undertaken to help judge the potential of these important regions, and the risks involved in intensifying use of them.

One of the major problems in land classification is to achieve adequate division of the medium class lands. It is not difficult to

[1] Kellogg (1961).
[2] Canada Land Inventory (1965).
[3] N.Z. Soil Conservation and Rivers Control Council (1969).
[4] Many other land classification schemes have been developed in other countries, some being referred to in MAFF (1974).

dentify the distinctive features that indicate whether a tract of
and is of especially high quality or suffers from such limitations
of use that it must be rated low, and in such cases allocation to
classes is fairly straightforward. The tendency is to delimit these
areas and to leave the remainder in medium grades which then
become too broad. This is the case with the A.D.A.S. Grade III,
which covers almost 50 per cent of the whole.[1]

Land Classification in Tropical Regions

The formulation of a land classification system for use in
tropical regions must obviously have regard to the fact that very
different conditions prevail from those in the temperate regions
where land classification has hitherto been most developed.
Systems used in temperate lands can give guidance to surveyors
in the tropics in general principles only. Certain factors will have
very different effects in their operation, e.g. increasing altitude is
generally detrimental in temperate regions but may be advantageous in the tropics. Nevertheless, though its effects will be
different, altitude can be used to make certain broad delineations,
and may be a good starting point once its effects are determined.
Vegetation is, of course, often a valuable index of climate and soil
and may sometimes be used as a short cut given adequate
knowledge of the habitat factors of species and associations
involved.

In the assessment of the productive capacity of soils it is
considered in the temperate lands that physical characteristics are
much more important than chemical status. This is largely
because phosphates, nitrogen, potash and lime can be applied
economically in large quantities and on a regular basis in
commercial farming for lucrative markets. This assumption must
be examined with caution before it is applied to any tropical
conditions, partly because of the very high rate of leaching typical
of the tropics and partly because application of fertilisers in most
tropical economies is of extremely limited practicability.

The rapid decline in fertility of tropical soils under cultivation
has been reported from many sources.[2] Decline is particularly
rapid in humid forest regions. In Ghana, trials for 8 years under a
continuous 2-year rotation of maize and cassava resulted, in the

---

[1] Weiers and Reid (1974), 1.
[2] Nye and Greenland (1960), summarise a number of examples.

absence of fertiliser effects, in the yields in the fourth cycle being only half those of the first cycle in both crops. The first cycle followed clearing of mature secondary forest.[1] Virgin land in Malaysia has been estimated to produce 2000, 1400 and 1000 kg of paddy per acre in successive years.[2] Applications of manure commonly have an effect lasting only for a few months as compared with two or three years in temperate regions. Hence, a practical classification of tropical lands may have to distinguish classes for which shifting cultivation or land rotation is necessary for physical reasons.

The importance that can be assigned to soil texture is probably not yet adequately evaluated, though Gourou[3] has written 'tropical agriculture is satisfied with the poorest soils, provided that they have a suitable texture, that is to say, that they are sufficiently friable'. This is a generalisation which cannot apply equally to all cultivated, tropical plants but may help in evaluating criteria for classification.

Considerable areas in the tropics are practically valueless agriculturally, because of lateritic crusts or rock outcrops. Over most of the remainder the soils are mediocre but rich deltaic lands such as those of the Song, Mekong, Chao Phraya, Irrawaddy Brahmaputra, Ganges and Indus are renewed by elements transported from mountains of sedimentary rocks. On the other hand the Niger has no such source of nutrients and its delta has poor soils.[4] Recent marine alluvium and recent basic volcanic ash are two other highly fertile groups of parent materials.

As in the temperate regions, it should not be difficult to establish classes for the types of lands of lowest fertility, bearing in mind that the minimum distinctions that will be of value must separate land types capable of improvement from those which have no significant potential. Laterites, and soils in which lateritic processes are far advanced must be regarded as among the worst. Gourou reports on many areas where lateritic soils are almost useless and records that in Madagascar relatively friable laterites have been transformed by irrigation only into a compact puddled clay, whereas water has enabled crops to be forced from the pure and almost sterile sand of the coast strip.

It is with the subdivision of the medium quality classes that the

[1] Nye and Stephens (1960), Chapter 7.   [3] Gourou (1966), 17.
[2] Grist (1953, 1959), 331.   [4] Gourou (1966), 27–28, 119.

greatest difficulty can be expected. This is especially likely in tropical conditions where there is a wide range of land types and where subsistence and commercial agriculture are often inter-mixed. In these circumstances, the need for a map which can be produced quickly for practical planning purposes may suggest priority for the type of classification which states the use for which each tract of land is suitable, without, at any rate in the reconnaissance stage, attempting to put a comparative label on it.

An example of this type of classification was that produced for British Honduras, now Belize. The land potential map was issued with soil maps and a comprehensive report.[1] Its approach to the problem of classification was to set out the following main categories:

Land adapted to forest use — protection forest, mahogany forest, pine forest.

Land adapted to agricultural use — orchard crops, long rotation pasture, short rotation pasture and arable crops, swamp rice, market gardening, bananas, sugar cane.

Problem soil which will become useful only if main drainage or desalinisation projects are expedited.

Subdivisions were made according to the particular crops or crop combinations recommended.

It is difficult to see how a land classification of either kind can be of value in the urgent task of increasing output unless it has regard to practical limitations imposed by institutional factors, land tenure, social habits, marketing opportunities, etc. It may be necessary for a classification to be multiple, e.g. a long-term optimum and a more immediate aim for a given area. This is well brought out in the report of the land use survey team in British Honduras. It was the conclusion of the survey that the highest production values per acre were likely to be achieved through forestry on a sustained yield basis, with the next best sustained return per acre from grass farming. The second conclusion is interesting, applying as it does to regions of over 4000 mm annual rainfall and high mean temperatures. But equally it was recog-nised that the forest resources and grass farming would take time to develop, and meanwhile other forms of production would have to be developed to meet the urgency of the economic

[1] Romney (ed.) (1959).

situation. It was suggested that for one or more decades special efforts should be put into producing rice, citrus fruits, beans and cacao since these could be sold in export markets to finance long-term developments. Soils ideal for these crops are not extensive enough to yield a worth-while volume of produce, so it was suggested that there would have to be some deliberate misuse of land for a time, using for these crops soils which would be more productive under well-managed grass.

The Punta Gorda sub-region in the southern area illustrated

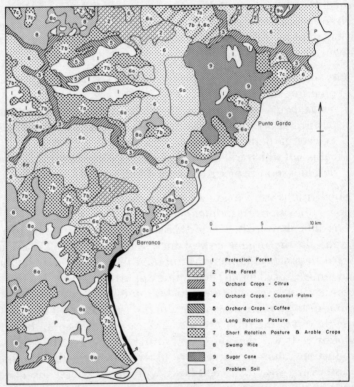

Figure 18. Potential land use, British Honduras. An extract from one of the maps in the survey report of 1959. Subdivision of classes: 6, Long rotation pasture with beef cattle; 6a, With dairy cattle; 7a, Leys with corn, beans; 7b, Leys with upland rice; 7c, Leys with swamp rice; 7d, Leys with cassava, pineapple; 8, Swamp rice, suitable for early development; 8a, Swamp rice, restricted by engineering problems; 9, Sugar cane; 9a, Sugar cane in rotation with grass.

this dilemma. Here it was suggested that the long-term plan should concentrate on cattle raising and this is shown on the example from the land potential map (Figure 18). The farmers would, however, need instruction on sowing pastures. Marketing arrangements would be necessary to give confidence in the future of the industry, but the immediate need was for a reliable cash crop which might not necessarily be retained in the economy of the district except as a subsistence crop. For this purpose upland rice was recommended for this region.

Immediately to the south is a region where bananas and pineapples have provided useful income in the past but have failed to find regular markets. Near Barranco, swamp rice is grown in the coastal depressions. In this region also it was recommended that attention should be given to the possibilities of establishing pastures for cattle raising. The construction of a wharf would facilitate exports of cattle, nuts, rice, etc., until which time attention should be concentrated on crops least likely to suffer damage through handling and transit. Potential ricelands could become very productive, given polders, and irrigation and engineering advice was called for in the report.

Unit-area Land Classification

Land classification schemes vary from reconnaissance to detailed field surveys, the former suffering from obvious doubts regarding applicability in detail, the latter from time and cost problems. To try to bridge the gap more effectively a unit-area method of land classification was developed for the Tennessee Valley Authority.[1] In this method the relevant factors of both physical and human environment are tabulated and numbered, and the numbers for each unit area of land are expressed in a fractional code. The denominator contains the digits representing the physical conditions, while the numerator shows the human data. A new unit-area is delimited whenever a change occurs in one or more of the items recorded.

The complete symbol for each land unit is in three parts — a Roman numeral, a short fraction and a long fraction, for example

$$III \; \frac{3}{4} \; \frac{2B233}{4122234} \, .$$

[1] U.S.A. National Resources Planning Board (1941), 119.

The Roman numeral shows the severity or absence of problems in the area, as revealed by the details that follow, in five classes; I problems insignificant; II problems not critical; III moderately critical; IV very critical; V so critical that a radical change in land use appears desirable.

The short fraction is a summary of the long fraction:

NUMERATOR

Areas classified on the basis of:

Class 1. EXCELLENT — Medium to large fields; little or no idle land; complete agricultural utilisation of the land under a regime especially well suited to the land.

Class 2. GOOD — Medium to large fields; little or no idle land; efficient agricultural use under a regime well suited to the land.

Class 3. MEDIUM — Small to medium fields; limited idle land; moderately efficient agricultural use under a regime moderately well suited to the land.

Class 4. POOR — Small to medium fields, in many cases interrupted; considerable idle land; low-grade, inefficient, or destructive agricultural use under a regime ill-suited to the land.

Class 5. VERY POOR — Very small, interrupted fields; excessive idle land; very low grade, inefficient, or destructive agricultural use under a regime in most cases entirely unsuited to the land.

DENOMINATOR

Areas classified on the basis of the present quality of the land for arable farming.

Class 1. EXCELLENT — Lands exceptionally well suited to intensive arable farming.

Class 2. GOOD — Lands well suited to arable farming and moderately suited to intensive forms of arable farming under proper land management.

Class 3. MEDIUM — Lands suited to less intensive forms of arable farming or to general farming under proper land management.

Class 4. POOR — Lands for the most part poorly suited to all forms of arable farming even under proper land management; in many cases best suited to grazing.

Class 5. VERY POOR — Lands not suited to either arable farming

TABLE 9

UNIT AREA LAND CLASSIFICATION CRITERIA FOR DELIMITING HOMOGENEOUS LAND USES

A. Major land uses, shown by the numerator of the long fraction

First Digit	Second Digit	Third Digit	Fourth Digit	Fifth Digit
Major Land Use	Agricultural Emphasis	Field Size	Amount of Idle Land	Quality of Farmsteads and Equipment
General farming	A. Corn	1. Large	1. Little	1. Excellent
Animal industry	G. Grain (small)	2. Medium	2. Limited	2. Good
Cash-crop farming	B. Beef cattle	3. Small	3. Considerable	3. Medium
Part-time farming	D. Dairying	4. Very small	4. Excessive	4. Poor
Subsistence farming	S. Sheep			5. Very poor
Forest land	H. Hogs			
Recreational area	M. Mules and/or horses			
Rural-village area	P. Poultry			
Urban area	T. Tobacco			
Manufacturing and mining areas	C. Cotton			
	W. Truck			
	O. Orchard			
	N. No emphasis			
	F. Forage			

B. Major physical conditions, shown by the denominator of the long fraction

First Digit	Second Digit	Third Digit	Fourth Digit
Slope	Drainage	Erosion	Stoniness
Relatively level	1. Thorough	1. Little or no observable erosion	1. Free from stones
Relatively level to undulating	2. Adequate	2. Little denudation by erosion	2. Moderately stony
Undulating to moderately hilly	3. Poor	3. Sheet erosion and ephemeral gullies	3. Stony
Hilly	4. Very poor	4. Excessive sheet erosion and gullying	4. Very stony
Steep	5. Excessive	5. Excessive gully erosion	

Fifth Digit	Sixth Digit	Seventh Digit
Rock Exposure	Soil Depth	Soil Fertility
Little or no rock exposure	1. Deep (180 cm or more)	1. Exceptionally fertile
Limited rock exposure	2. Moderately deep (90–180 cm)	2. Fertile
Considerable rock exposure	3. Shallow (30–90 cm)	3. Moderately fertile
Excessive rock exposure	4. Very shallow (less than 30 cm)	4. Low in fertility
Rock exposure dominant		5. Very low in fertility

Summary after Jacks (1946): for details see U.S.A. National Resources Planning Board (1941), 121.

or grazing; in most cases suited to forest production, recreation, etc.; in some cases, waste land.

The criteria used in the long fraction are given in Table 9.

The suitability of this method of classification and mapping depends on availability of facilities for detailed field work, aerial photographs or detailed base maps and sufficient staff. The classification is not, however, well suited to distinguishing potential from existing productivity and the varying requirements for different uses are not recognised adequately.

Quantitative Classification

In the examples of land classification so far discussed there is no attempt to relate one class to another quantitatively, but many such attempts have been made to devise an objective and quantitative form of land classification. Jacks[1] says 'The methods used may be classed generally as inductive (e.g. by adding or otherwise integrating "marks" awarded to certain properties of the land or soil that influence productivity) or deductive (e.g. deduced from yield data), but many methods in actual use are a combination of both types.'

Jacks summarises several methods which indicate the range of approaches. The U.S. Soil Survey *productivity ratings* illustrate the deductive type of classification. An example is given as follows:

Soil type: Miami loam

Corn		Wheat		Rye		Alfalfa		Sugar Beet		Productivity grade	
A	B	A	B	A	B	A	B	A	B	A	B
70	90	70	100	80	100	70	90	60	70	3	1

Space is not available here to discuss the scheme in detail but, briefly, the A columns show the yields of crops with the common practices of management in the area, and the B columns those with the best practices, as percentages of standard yields. Standard yields represent average yields 'on the more extensive and widely developed soils of the regions in the United States in which the crop is a principal product.'[2] Thus, the standard for corn is 50 bushels per acre, and in the example given the average yield under average management is 35 bushels and under the

---

[1] Jacks (1946), 68. [2] Ableiter (1940).

best management, 45 bushels. The 'productivity grade' is obtained by a simple percentage weighting of the crop ratings according to each crop's local importance. Soils with a weighted average between 100 and 90 are graded 1, between 90 and 80, 2, and so on.

The 'Storie index' is an inductive method which has been used in other countries as well as in the U.S.A. It is based on the soil profile.[1] The main characteristics of the profile are expressed in three 'factors', one being the surface texture. Each is quoted as an estimated percentage of the optimum conditions of that factor for plant growth, and the factors are then multiplied to give the rating as a percentage of the possible 'score'. Multiplying instead of adding the figures for the characteristics enables any one very detrimental factor to bring down the index for the soil as a whole,

e.g. $100 \times 100 \times 10 = $ ratio 10, i.e. $\left(\dfrac{100,000}{1,000,000}\right)$ whereas $100 + 100 + 10 = $ ratio 70, i.e. $\left(\dfrac{210}{300}\right)$ by addition.

A much more extensive use of points systems is the basis for the German soil classification and land valuation procedure. This was first adopted in 1934 and continues in use in the German Federal Republic following post-war revisions and also forms the basis of classification schemes adopted in some Eastern European countries. It is a multi-stage classification involving (a) the identification and description of the soil, and (b) the establishment of the yield potential, taking into consideration soil conditions, topography and climatic conditions. In the final assessment of farm valuation, adjustments are made for conditions affecting economic use, such as location, distance of fields from buildings and difficulties in use of machinery owing to farm characteristics.[2]

In the evaluation of arable and pasture lands, different criteria are used, with more emphasis on climatic and water conditions in the latter case. (Later work recognising the greatly differing requirements of different arable crops as well as in flexibility of land use under progressive management suggests that this approach might be too rigid.) Texture, profile and structure are assessed and points allocated relative to the maximum of 100 points allocated to the soil considered to be the best in the country. Reductions or bonuses are given for deviations from

[1] Storie (1933, 1948).   [2] Weiers and Reid (1974).

norms in topography, climatic conditions, farm size and general farming conditions. Thus far the classification is concerned mainly with soil types, and, to a lesser extent, with climatic conditions but characteristics of the functional farm unit are also assessed in the ensuing standard valuation. The soil classification and standard valuation are used for a wide range of purposes. It is claimed that agricultural rents are closely related to the soil classification figures and that nearly all advertisements for farm sales show at least the average results of the soil classification.[1]

It will be evident that although quantitative, such schemes have not been able to avoid a degree of subjectivity in the allocation of the points, or in deciding what constitutes normal management, optimum conditions, etc. It is possible for a greater degree of comparability to be obtained than through the type of classification previously discussed, but additional caution is necessary lest the presence of a statistical valuation leads to an uncritical acceptance of its validity.

Land Classification in the Soviet Union

One of the most comprehensive of land classification schemes was put forward by Zvorykin. The work would begin with the physical characteristics of the land, including relief, slope, erosion, soil properties and climatic data including dates when the land is ready for work in the spring, frost-free periods and flooding risks. Established land use would be mapped and examined critically, and data collected and processed on crop yields, and on the labour and cost required to grow crops on various kinds of land with different natural properties.[2] The object is to grade the land into a variety of types and sub-types suitable for a given kind of land use and for growing various crops or groups of crops. They are thus similar to the land potential classes recognised in the British Honduras survey referred to above. The British Honduras survey relied, it would seem, to a greater extent on the interpretation of the soil pattern. This was probably necessary in the conditions in which this survey was working, but Zvorykin argues that a solution of the problems of classification and land appraisal is hindered by undue emphasis on soil types. For example, he cites three soil types of which a podzolised chernozem and a weakly-podzolised cher-

---

[1] Weiers and Reid (1974), 35.   [2] Zvorykin (1963).

nozem belong together pedologically, while the dark-grey soil belongs to a different type. He found, however, that the dark-grey soil and the podzolised chernozem fitted in production potential into the same sub-group, whereas the weakly-leached chernozem belongs to another sub-group.

Soil survey may in fact produce a complexity of detail which is unnecessary and possibly even misleading from the point of view of land potential, but if reliable values can be attached to each unit of the soil classification, then, of course, the soil map will be of great assistance to the land classification. Similarly with relief and landforms, of which practical data such as slope and micro-relief characteristics are needed rather than genetic interpretations.

Coverage of an immense area such as the Soviet Union in such detail is a huge task but progress along these lines is to be seen in the land classification maps produced by teams such as that led by Zvorykin as far afield from their Moscow base as eastern Siberia.

Current Progress in Detailed Analysis

The amount of work involved in detailed farm productivity in relation to soil variations may be illustrated by work in Northern Ireland. Criticising, in particular, reliance on soil profile characteristics to indicate land quality, Cruickshank and Armstrong note that no effort has been made by most classifiers of land potential to test the influence of soil properties measured in the laboratory on measured agricultural productivity. Even if relationships are firmly established, they point out, interpretation must proceed with caution.[1] Effects will vary according to the crops, as for example, with lime status, an increase in which produces improved yields of most crops, but has adverse effects on potatoes, an important crop in Northern Ireland. Because of the lack of detailed soil maps it was necessary to prepare a special map of the Roe Valley, which had been chosen for the investigation. Economic data were obtained from a stratified random sample of ten farms from each of seven soil series, all farms having to meet certain stipulations on size, location, and other features, and soil samples were also collected on the farms. Multiple regression analysis showed that the significant soil properties explained between 35 and 43 per cent of the variation in gross margins, but the spatial variation of the significant

[1] Cruickshank and Armstrong (1971).

properties did not coincide with the mapped soil series units. The evidence of the five equations used, however, suggested that separate maps of individual soil properties would be required if one wished to show significant variations in soils for different farm enterprises. This work confirms views already expressed that soil survey is only one element, albeit a very important one, in land classification, which is essentially a task for co-ordinated, multidisciplinary effort.

Crop Protection and Conservation

The emphasis in this chapter has been on assessment of land for increasing production, but if the full value of the output at any level is to be enjoyed by the people for whom it is intended it is necessary to minimise loss of crops both when growing in the fields and when harvested. All the planning that can precede farming operations and all the care in the actual cultivation processes, the inputs of land, seed, machinery, fertilisers and energy of all forms can be lost to the depredations of disease, insects and rodents. Widespread pathological research has not been accompanied by any substantial amount of work in the spatial aspects of crop diseases or other forms of loss, which is perhaps the more surprising considering the growth of medical geography.

Crop protection by chemical application is a controversial aspect of modern agriculture but it appears to be essential if yields are to be maximised, because in the intensifying battle for survival which man will face as the earth becomes more crowded, the control of insects and pests is of fundamental importance. Control of the tsetse fly alone might make available for livestock production some 7 million square km of the African continent, sub-tropical grainlands are threatened by locusts and less well-known pests such as the weaver bird which is a major problem in the Sudan, while cotton in the Sudan and elsewhere in Africa is attacked by *Heliothis armigera* larvae. There is no possibility here of extending the sequence to show the vast range of problems that beset agriculture from pests, diseases and weeds but their extent and the success and failure of control methods are still inadequately studied and there is scope for geographers to join other scientists in this work.

CHAPTER 11

Conclusions

Agricultural geography has evolved as an academic discipline, initially descriptive, but developing its own analytical procedures as its practitioners refined their methods of statistical and cartographic representation of relevant data in a constant search for greater objectivity in their depiction of the agricultural landscape. The resulting works have included many meticulous statements of observed facts and subsequent analysis which, using a precisely delimited spatial framework, could serve as models for public servants and planners concerned with producing their own surveys for specific purposes, as well as providing a mass of valuable information, processed and presented with the support of maps which allow of little ambiguity in data-display. Since the mid-'fifties immense progress has been made in the amount of data that can be handled and the speed and effectiveness with which it can be presented in a form in which it can be put to practical use, computers playing a major part in these advances.

Professional geographers and other research workers studying the landscape for its own intrinsic interest have tended to be content with developing concepts and methods for the better analysis of problems of an academic nature. Though, increasingly, their work can be seen to have a bearing on practical issues and policies, particularly of such problems as farm size and layout, accessibility, marketing, competition for land and labour and productivity in advanced societies and in the countries of the Third World, such applications have usually been of secondary consideration when originating their work. This orientation is appropriate in that academic research must be concerned with fundamental truths and their revelation, the publication and discussion of which should not be hindered by or modified by political, sectional or other prejudiced and transitory interests.

253

Nevertheless, the world is beset, as we approach the last two decades of this millenium, with problems of a severity which threaten the future of mankind, and, indeed, of other forms of life, not anticipated only a few decades ago. After the Second World War there was a widespread feeling of optimism that, with total war having become too awful to contemplate (except in terms of its prevention) and technical and economic problems of development, hitherto insoluble, being overcome, the possibilities for human progress were unlimited. The release of vast areas from colonial bondage heightened appreciation of their backwardness, but with the concept of the Third World as a great area of development with aid from both capitalist and communist advanced economies, hope for the future was paralleled by confidence that the problems would be overcome. The 'green revolution' was a manifestation of such progress in the sphere of agriculture. During the 1960s, however, the image began to become tarnished as politically inspired conflicts intensified, trading blocs hardened, international agreements encountered more and more handicaps, the gap between rich and poor countries widened, and the population explosion was seen as a growing threat to economic and social standards. There were, and still are, some who scorned 'neo-Malthusian' arguments that world population would outstrip food supplies, but the great majority of thinking people became increasingly alarmed at the prospects. The attention of the wealthier nations turned from aiding the Third World to consolidating their own position, and the reasoned cry for 'trade not aid' was met in all too many areas by raised rather than lowered barriers to free interchange. As the 'seventies advanced, disillusionment spread as bitterness in some of the de-colonialised lands turned to violent self-assertion while old disputes seemed no nearer settlement. The crisis in energy supplies initiated by the Arab owners of the vital Middle East oil resources who thought, not without some justification, that they were being too little rewarded for their supply to world markets, caused widespread dislocation of trading patterns and depression of markets which hit the oil-poor countries of the Third World more than the capitalists against whom the measures were primarily aimed. It was also then realised that productivity in agriculture could not be raised limitlessly through application of more energy, because that too was becoming relatively scarce and expensive. The inability of rich and poor countries alike to control their currency problems and match consumption to production further deepened the world's troubles. The more successful producers and traders became more conscious of their need to hold and extend their markets and even more desirous of amassing huge profits and the less successful increased their efforts to emulate those who had shown the way to wealth and power.

This sad catalogue holds many messages for practical agriculturalists and for academic agricultural geographers. Those whose concern is simply to produce food by honest endeavour and to market it at fair prices find themselves ensnared in a web of political and commercial intrigue and exploitation and need all the help they can get from those whose involvement is less immediate but whose perspective may be valuable for long-term planning and adaptation. It is not easy to guide research into selected channels without endangering academic freedom. Research must remain untrammelled by sectarian, political or commercial bonds, for scholarship will not flourish unless left free. If a research worker is convinced that there will be benefits if he embarks on an investigation into the origins of groundnut cultivation he should be free to do so, and not only because his research may chance to reveal some information that may lead to more successful cultivation of the crop, nor should he be discouraged from analysing and 'merely' describing an existing distribution even if he does not appear to be developing new research tools — though these may appear as the work proceeds.

This argument does not invalidate a personal view that agricultural geography is a discipline (or, if one prefers, a branch of a discipline) that can and must do more to help resolve world food and raw material problems. Without sacrificing any academic integrity more of the research now emerging could include a positive element of intention to pursue and reveal knowledge likely to be of benefit to producers and planners. Geographers believe in the importance of their work in analysing landscapes, spatial distributions, locational factors, etc., as do other scientists and workers in many disciplines, but they do not always take as much pride as they might in orientating their efforts to positive results beyond the purely academic. It is the writer's hope that this further consideration will become more common as the discipline advances in complexity and sophistication, so that its findings may be more generally available, understood and applied. Any contribution that agricultural geography can make towards providing means of feeding adequately the world's growing population, is ultimately more important than its academic justification.

Each type of agriculture has its own problems, both physical and cultural. Systems of subsistence agriculture are usually carried on with few technical aids and little scientific knowledge. Their disturbance is fraught with danger and should be encouraged only after the most careful study, but modest investment in wells and other improvements in water supply, improved tools including simple machinery and some better seeds and livestock breeding can pay immense dividends. Plantation agriculture, in the form that Europeans have developed it in tropical lands over four centuries, seems likely for political reasons to be doomed, at least in some countries where it is still a major source of economic strength, but alternative forms of organisation are being found to take its place. State-operated and collective farms have not measured up to the hopes of Soviet planners, but compromises between these forms and private enterprise practised by a modernised peasantry may yet prove a solution in some areas. Eastern European, Chinese, Cuban, Israeli and other experiments must be closely watched for the lessons they can convey to other societies.

European countries which are advanced technologically are burdened with excessive numbers of small, fragmented and inefficient farms. Here, too, the competition for land is acute as the swelling population demands more and more land for housing, factories, roads, playgrounds, water catchment and other uses, and a worsening of these problems must be expected as the world population rises. Even in North America, Australia and New Zealand, where land is still relatively plentiful, there are problems of farm size and shape and of technical and climatic difficulties as well as hindrances to the disposal of produce in conditions of impeded international trade. If, furthermore, the tendency of people to congregate in urban and suburban areas continues (and there is no sign of any reversal of this trend) there will be a continuous reduction in the proportion of those who are actual food producers. Eventually, unless production can be continuously increased per hectare and per man, there must be a crisis situation.

Because of the surpluses which have been built up in some areas, mainly in the E.E.C. and the U.S.A., the risk that the world may experience an acute food shortage has been obscured, but even in this situation many scientists have wisely given consideration to the limits on food and raw material production. Ultimately, all crop production results from the conversion of solar energy through photosynthesis, and the availability of warmth and water and soil nutrients from natural and artificially adjusted cycles, and animal production depends on vegetation, whether natural or cultivated. Many attempts have been made to assess the ultimate potential of the earth to produce food and raw materials and estimates vary considerably. One such calculation of the absolute maximum production theoretically possible on the land surface of our planet suggested that, whereas approximately 1400 million hectares are now cultivated, agriculture could be extended to more than twice this area.[1] Even this total would give no room for complacency with current and projected increases in population, but in fact the theoretically available area must be much reduced because of restrictions imposed by local physical conditions and alternative forms of land use. Production could, however, be increased very greatly. Theoretically, the present grain growing area of the world could produce thirty times its present crops,[2] but in practice the unreliability of weather, shortage of water for irrigation, pests, diseases of crops and other limitations would prevent such limits being attained.

With production further retarded by economic and institutional obstacles while only a minority of the world's population enjoys a satisfactory diet, the applications of agricultural geography must be added to those of other sciences. Geographical methods can be used advantageously in analysis of farm units, patterns of cropping and livestock, intensity of usage, diffusion of techniques, effects of changed methods, land tenure patterns, transport networks, effects of marketing schemes and subsidies, and innumerable other aspects of agriculture. Studies of land use and land potential lead to a more comprehensive and more valuable inventory of agricultural resources and their most

---

[1] Meadows *et al.* (1972), Buringh *et al.* (1975).
[2] Buringh *et al.* (1975), 1.

effective employment. Geographers have no monopoly of any of these approaches and techniques; and advances will be made all the more rapidly if there is appreciation on all sides of the interdependence of the several disciplines that contribute to the solution of any geographical problem.

It is hoped that the reader will have been stimulated already to refer to some of the works cited in the text. Many of these are 'secondary' sources particularly valuable for taking studies a stage further while being more widely available than many of the papers and monographs that provide for a thorough understanding of the subject. References and bibliographies in the works quoted open up a vast literature in several languages.

METRIC EQUIVALENTS

1 millimetre	= 0·0394 inch
1 centimetre	= 0·3937 inch
1 metre	= 3·281 feet, 1·0936 yards
1 kilometre	= 0·6214 mile
1 decare	= 1196 square yards or 0·2474 acre
1 hectare	= 2·4736 acres
1 square kilometre = 100 hectares	= 0·3861 square mile
1 kilogram	= 2·2046 pounds
1 tonne = 1000 kilograms	= 0·984 long ton
1 litre	= 1·76 pints, 0·22 imp. gallon, 0·264 U.S. gallon

Bibliography

Ableiter, J. K. (1940) 'Productivity ratings of soil types'. *Missouri Agric. Expt. Sta. Bull.*, 421.

Academy of Sciences of the U.S.S.R. *See* U.S.S.R.

Anderson, J. (1967) 'A historical-geographical perspective on Khrushchev's corn program', in Karcz, J. F. (ed.), *Soviet and East European agriculture.* Berkeley and Los Angeles.

Anuchin, V. A. (1961) 'On the subject of economic geography', *Geografiya i khozyaystvo*, trans. in *Soviet Geography: Review and Translation*, Vol. 2, No. 3, 26–43.

Avery, B. W. (1973) 'Soil classification in the soil survey of England and Wales'. *Jnl. Soil Science*, Vol. 24, 324–338.

Baker, O. E. (1925) 'The potential supply of wheat'. *Economic Geography*, Vol. 1, 15–52. (*See also series of articles by this author in subsequent issues of this journal.*)

Beavington, F. (1963) 'The change to more extensive methods in market gardening in Bedfordshire'. *Trans. Inst. British Geographers*, No. 33, 89–100.

Bell, T. S. (1974) 'Remote sensing for the identification of crop and crop diseases' in Barrett, E. C. and Curtiss, L. F., *Environmental remote sensing*, London, 153–166.

Beresford, M. W., and St. Joseph, J. K. S. (1958) *Medieval England, an aerial survey.* Cambridge.

Berry, B. J. L. (1958) 'A note concerning methods of classification'. *Annals Assoc. American Geographers*, Vol. 48, 300–303.

Best, R. H. (1960) *The major land uses of Great Britain.* Wye College.

Best, R. H., and Coppock, J. T. (1962) *The changing use of land in Britain.* London.

Best, R. H., and Ward, J. T. (1956) *The garden controversy.* Wye College.

Bibby, J. S., and Mackney, D. (1969) *Land use capability classification*, Soil Survey Technical Monograph, No. 1, Rothamsted.

Birch, J. W. (1954) 'Observations on the delimitation of farming-type regions, with special reference to the Isle of Man'. *Trans. Inst. British Geographers*, Vol. 20, 141–158.

Blaikie, P. M. (1971) 'Spatial organisation of agriculture in some north Indian villages'. *Trans. Inst. British Geographers*, No. 52, 1–40; No. 53, 15–30.

Board, C. (1968) 'Land use surveys: principles and practice', in *Land use and resources: studies in applied geography*, Inst. British Geographers, Special Publication No. 1, 29–41.

Bowler, I. R. (1976) 'Spatial responses to agricultural subsidies in England and Wales'. *Area*, Vol. 8, 225–229.

Bridges, E. M. (1970) *World soils.* London.

Brookfield, H. C., and Brown, Paula (1963) *Struggle for land. Agriculture and group territories among the Chimbu of the New Guinea Highlands.* Melbourne.

Brookfield, H. (1973) *The Pacific in transition.* London.

Buchanan, R. O. (1935) *The pastoral industries of New Zealand*, Inst. British Geographers. Pubn. No. 2.

Buchanan, R. O. (1951) 'Approach to economic geography'. *Indian Geographical Journal*, Silver Jubilee Souvenir Volume, 1–8.

Buchanan, R. O. (1959) 'Some reflections on agricultural geography'. *Geography*, Vol. 44, 1–13.

Budyko, M. I. (1956) *Teplovoy balans zemnoy poverkhnosti*, Leningrad, trans. by N. A. Stepanova, *The heat balance of the earth's surface*, U.S. Dept. of commerce, Washington, 1958.

Bunge, W. (1962) *Theoretical geography.* Lund.

Buringh, P., van Heemst, H. D. J., and Staring, G. J. (1975) *Computation of the absolute maximum food production of the world.* Agricultural University, Wageningen.

Canada Land Inventory (1965) *Soil capability classification for agriculture*, Report No. 2, Dept. of Forestry and Rural Development.

Chakravarti, A. K. (1976) 'The impact of the High-Yielding Varieties Program on foodgrain production in India. *Canadian Geographer*, Vol. 20, 199–223.

Chisholm, M. (1962) *Rural settlement and land use.* London.

Chisholm, M. (1964) 'Problems in the classification and use of farming-type regions'. *Trans. Inst. British Geographers*, No. 35, 91–103.

Chisholm, M. (1966) *Geography and economics.* London, 2nd ed. 1970.

Clark, C. (1973) *The value of agricultural land.* Oxford.

Clark, C., and Haswell, M. R. (1964) *The economics of subsistence agriculture.* London.

Clarke, J. G. D. (1952) *Prehistoric Europe: the economic basis.* London.

Clout, H. (1972) *Rural geography: an introductory survey.* Oxford.

Coleman, A., and Maggs, K. R. A. (1961) *Land use survey handbook.* (2nd ed.) Isle of Thanet Geographical Association.

Coppock, J. T. (1964a) 'Crop, livestock and enterprise combinations in England and Wales'. *Economic Geography*, Vol. 40, 65–81.

Coppock, J. T. (1964b) *Agricultural atlas of England and Wales.* London.

Coppock, J. T. (1971) *An agricultural geography of Great Britain.* London.

Crossley, J. C. (1976) 'The location of beef processing'. *Annals Assoc. American Geographers*, Vol. 66, 60–75.

Cruickshank, J. G., and Armstrong, W. J. (1971) 'Soil and agricultural land classification in Co. Londonderry'. *Trans. Inst. British Geographers*, No. 53, 79–94.

Cruickshank, J. G. (1973) *Soil geography*. Newton Abbot.

Cumberland, K. B., and Fox, J. W. (1962) *New Zealand, a regional view*. Christchurch (revised reprint, 1970).

Curry, L. (1962) 'The climatic resources of intensive grassland farming: the Waikato. New Zealand'. *Geographical Review*, Vol. 52, 174–194.

Curry, L. (1963) 'Regional variation in the seasonal programming of livestock farms in New Zealand'. *Economic Geography*, Vol. 39, 95–118.

Curtis, L. F., Courtney, F. M., and Trudgill, S. (1976) *Soils in the British Isles*. London and New York.

Darling, F. F. (ed.) (1955) *West Highland Survey*. London.

Davey, B., Josling, T. E., and McFarquhar, A. (eds) (1976) *Agriculture and the State: British policy in a world context*. London.

de Schlippe, Pierre. *See* Schlippe.

Devon Commission (1847–48) *Digest of evidence taken before H.M. Commissioners of Enquiry into the state of the law and practice in respect of the occupation of land in Ireland* (2 vols.). H.M.S.O. Dublin.

Digby, M. (1963) *Co-operative land use, the challenge to traditional co-operation*. Oxford.

Donkin, R. A. (1963) 'The Cistercian Order in medieval England: some conclusions'. *Trans. Inst. British Geographers*, No. 33, 181–198.

Duckham, A. N., and Masefield, G. B. (1971) *Farming systems of the world*. London.

Duley, F. L., and Hays, O. E. (1932) 'The effect of the degree of slope on runoff and soil erosion'. *Jnl. Agricultural Research*, Vol. 45, 349–360.

Duncan, J. S. (1962) 'The land for the people: land settlement and rural population movements, 1886–1906', in *Land and livelihood, geographical essays in honour of George Jobberns* (ed. M. McCaskill), 170–190. N.Z. Geog. Soc., Christchurch.

Dunn, E. S. (1954) *The location of agricultural production*. Gainesville.

Eadie, J., and Cunningham, J. M. M. (1971) 'Efficiency of hill sheep production systems', in Wareing, P. F., and Cooper, J. P. (1971) (eds), *Potential crop production*, 239–249.

Elliott, F. F. (1933) *Types of farming in the United States*. U.S. Govt. Printing Office.

Ellison, W. (1953) *Marginal land in Britain*. London.

Evans, E. Estyn (1956) 'The ecology of peasant life in western Europe', in *Man's role in changing the face of the earth* (ed. W. L. Thomas). Chicago, 217–239.

Eyre, S. R. (1963) *Vegetation and soils; a world picture*. London, 2nd ed. 1968.

Farmer, B. H. (1960) 'On not controlling subdivision in paddy lands', *Trans. Inst. British Geographers*, No. 28, 225–235.

Faucher, D. (1949) *Geographie agraire: types de cultures*. Paris.

Fielding, G. J. (1964) 'The Los Angeles milkshed: a study of the political factor in agriculture'. *Geographical Review*, Vol. 54, 1–12.

Fielding, G. J. (1965) 'The role of government in New Zealand wheat growing'. *Annals Assoc. American Geographers*, Vol. 55, 87–97.

Forde, C. D. (1934) *Habitat, economy and society*. London.

Fortes, M., Steel, R. W., and Ady, P. (1948) 'Ashanti survey, 1945–46: an experiment in social research'. *Geographical Jnl.*, Vol. 110, 149–179.

Franklin, S. H. (1962) 'Reflections on the peasantry'. *Pacific Viewpoint*, Vol. 3, 1–26.

Franklin, S. H. (1971) *Rural societies*. London.

Fullerton, B. (1954) 'The northern margin of grain production in Sweden in the twentieth century'. *Trans. Inst. British Geographers*, No. 20, 181–191.

Garnett, A. (1937) *Insolation and relief*, Inst. British Geographers. London.

Garrad, G. H. (1954) *A survey of the agriculture of Kent*. London.

George, P. (1962) *L'U.R.S.S.* (2nd ed.). Paris.

Gourou, P. (1966) *The tropical world* (4th ed.).

Green, F. H. W. (1976) 'Recent changes in land use and treatment'. *Geographical Journal*, Vol. 142, 12–26.

Gregor, H. F. (1963) 'Regional hierarchies in California agricultural production: 1939–1954'. *Annals Assoc. American Geographers*, Vol. 53, 27–37.

Gregor, H. F. (1965) 'The changing plantation'. *Annals Assoc. American Geographers*, Vol. 55, 221–238.

Gregor, H. F. (1970) *Geography of agriculture: themes in research*. Englewood Cliffs, N.J.

Gregory, S. (1954) 'Accumulated temperature maps of the British Isles'. *Trans. Inst. British Geographers*, No. 20, 59–73.

Grigg, D. (1965) 'The logic of regional systems'. *Annals Assoc. American Geographers*, Vol. 55, 465–491.

Grigg, D. (1969) 'The agricultural regions of the world: review and reflections'. *Economic Geography*, Vol. 45, 95–132.

Grigg, D. (1970) *The harsh lands, a study in agricultural development*. London.

Grigg, D. (1974) *The agricultural systems of the world, an evolutionary approach*. London.

Grist, D. H. (1953) *Rice*. London, 2nd ed. 1959.

Grotewald, A. (1959) 'Von Thünen in retrospect'. *Economic Geography*, Vol. 35, 346–355.

Haggett, P. (1965) *Locational analysis in human geography*. London.

Hahn, E. (1892) 'Die Wirtschaftsformen der Erde'. *Petermanns Mitteilungen*, Vol. 38, 8–12.

Hall, P. (ed.) (1966) *Von Thünen's Isolated State*. Oxford. Translation by Carla M. Wartenberg.

Halstead, C. A. (1958) 'The climate of the Glasgow region', In *The Glasgow region: a general survey* (eds R. Miller and J. Tivy). Glasgow, 62–72.

Harris, D. R. (1969) 'Agricultural systems, ecosystems, and the origins of agriculture' in Ucko, P. J., and Dimbleby, G. W. (eds), *The domestication and exploitation of plants and animals*, 3–15.

Hartshorne, R., and Dicken, S. N. (1935) 'A classification of the agricultural regions of Europe and North America on a uniform statistical basis'. *Annals Assoc. American Geographers*, Vol. 25, 99–120.

Hartshorne, R. (1939) *The nature of geography*. Lancaster, Pa.

Hartshorne, R. (1959) *Perspective on the nature of geography*. Chicago.

Haystead, L., and Fite, G. C. (1955) *The agricultural regions of the United States*. Norman, Oklahoma.

Heichelheim, F. M. (1956) 'Effects of classical antiquity on the land', in *Man's role in changing the face of the earth* (ed. W. L. Thomas). Chicago, 165–182.

Henderson, J. M. (1957) 'The utilisation of agricultural land: a regional approach'. *Papers and Proceedings, Regional Science Association*, Vol. 3, 99–117.

Higbee, E. (1958) *American agriculture*. New York.

Hogg, W. H. (1967) *Atlas of long-term irrigation needs for England and Wales*. Ministry of Agriculture, Fisheries and Food.

Hogg, W. H. (1970) 'Basic frost, irrigation and degree-day data for planning purposes', in *Weather economics* (ed. J. A. Taylor), 27–43.

Hoover, E. M. (1948) *The location of economic activity*. New York.

Howell, J. P. (1925) *Agricultural atlas of England and Wales*. Southampton.

Hoyle, B. S. (ed.) (1973) *Transport and development*. London.

Hutchinson, Sir Joseph (1965) *Essays on crop plant evolution*. Cambridge.

International Geographical Union (1952) *Report of the Commission on World Land Use Survey (1949–1952)*. Worcester, U.S.A.

International Institute for Land Reclamation and Improvement (1959) *Land consolidation in Europe*, prepared by E. H. Jacoby (F.A.O.). Wageningen.

Isard, W. (1956) *Location and the space economy*. Cambridge, Mass.

Jacoby, E. H. *See* International Institute for Land Reclamation and Improvement.

Jacks, G. V. (1946) *Land classification for land use planning*. Imperial Bureau of Soil Science, Harpenden.

Jackson, B., Barnard, C., and Sturrock, F. (1963) *The pattern of farming in the eastern counties*. Farm Economics Branch, Cambridge.

Jackson, W. A. D. (ed.) (1971) *Agrarian policies and problems in Communist and non-Communist countries*. Seattle and London.

James, O. G. (1971) 'Fertilizer application — large scale operations in New Zealand'. *Proc. Fourth International Agricultural Aviation Congress*, 1969. Wageningen, 496–500.

Johnson, D. Gale (1973) *World agriculture in disarray*. London.

Jonasson, Olof (1925) 'The agricultural regions of Europe'. *Economic Geography*, Vol. 1, 277–314.

Jones, W. D. (1930) 'Ratios and isopleth maps in regional investigations of agricultural land occupance'. *Annals Assoc. American Geographers*, Vol. 20, 177–195.

Jordan, T. G. (1969) 'The origin of Anglo-American cattle ranching in Texas; a documentation of diffusion from the Lower South'. *Economic Geography*, Vol. 45, 63–87.

Kawachi, K. (1959) 'On a method of classifying world agricultural regions'. *Proc. I.G.U. Regional Conference Japan*, Tokyo, 355–356.

Kellogg, C. E. (1961) *Soil interpretation in the soil survey*. Washington.

Kohnke, H., and Bertrand, A. R. (1959) *Soil conservation*. New York.

Konstantinov, O. A. (1962) 'Economic geography', in *Soviet geography, accomplishments and tasks*, trans. by L. Ecker. American Geographical Society, New York.

Kostrowicki, J. (1974) *The typology of world agriculture. Principles, methods and model types*. International Geographical Union, Warsaw.

Kostrowicki, J. (1976) 'Agricultural typology as a tool in planning spatial organisation of agriculture — I.G.U. Commission on Agricultural Typology'. *Geoforum*, Vol. 7, 241–250.

Lewthwaite, G. R. (1966) 'Environmentalism and determinism: a search for clarification'. *Annals Assoc. American Geographers*, Vol. 56, 1–23.

Lösch, A. (1940) *The economics of location*, Jena, trans. from revised edition, 1954. New Haven.

McCarty, H. H., and Lindberg, J. B. (1966) *A preface to economic geography*. Englewood Cliffs, N.J.

McClintock, A. H. (ed.) (1960) *A descriptive atlas of New Zealand*. Wellington.

McHugh, B. J. (1963) 'County Tyrone', in *Land use in Northern Ireland* (ed. L. Symons). London, 252–266.

Marshall, T. J. (1947) *Mechanical composition of soil in relation to field descriptions of texture*. Council for Scientific and Industrial Research, Melbourne.

Maruta, S. (1956) *Memoirs of the Faculty of Agriculture*. Kagoshima University, Japan.

Maude, A. (1970) 'Shifting cultivation and population growth in Tonga'. *Jnl. Trop. Geog.*, Vol. 31, 57–64.

Meadows, D. L., Randers, J., and Behrens, W. (1972) *The limits to growth*.

Messer, M. (1932) *Agricultural atlas of England and Wales*, 2nd ed.

Meteorological Office M.O. Form 3300 (1928) *Tables for the evaluation of daily values of accumulated temperature above and below 42 degrees F. from daily values of maximum and minimum temperature.* H.M.S.O. London.

Ministry of Agriculture, Fisheries and Food (1966) *Agricultural land classification,* ALS Tech. Report No. 11.

Ministry of Agriculture, Fisheries and Food (1974) *Land capability classification,* Tech. Bull. No. 30.

Morgan, W. B., and Munton, R. J. C. (1971) *Agricultural geography.* London.

Murray, J. (1970) *The first European agriculture.* Edinburgh.

Narr, K. J. (1956) 'Early food producing populations', in *Man's role in changing the face of the earth* (ed. W. L. Thomas). Chicago, 134–151.

New Zealand Department of Scientific and Industrial Research (1962) *Soil survey method,* Soil Bureau bulletin No. 25 by N. H. Taylor and I. J. Pohlen. Wellington.

New Zealand Department of Statistics, *Agricultural Statistics* (annual).

New Zealand Ministry of Agriculture and Fisheries (1974) *New Zealand Agriculture.* Wellington.

New Zealand Soil Conservation and Rivers Control Council (1969) *Land use capability survey handbook.* Wellington.

Nye, P. H., and Greenland, D. J. (1960) *The soil under shifting cultivation.* Commonwealth Agricultural Bureaux, Farnham Royal.

Nye, P. H., and Stephens, D. (1960) *Agriculture and land-use in Ghana.*

Ochse, J. J., Soule, M. J. (Jr.), Dijkman, M. J., and Wehlburg, C. (1961) *Tropical and subtropical agriculture,* 2 vols. New York.

Ooi Jin-bee (1961) 'The rubber industry of the Federation of Malaya'. *Jnl. Tropical Geography,* Vol. 15, 46–65.

Ooi Jin-bee (1963) *Land, people and economy in Malaya.* London.

Orwin, C., and Whetham, E. (1964) *History of British Agriculture.* London.

Peattie, R. (1936) *Mountain geography.* Cambridge, Mass.

Pedler, F. J. (1955) *Economic geography of West Africa.* London.

Perry, P. J. (ed.) (1973) *British agriculture 1875–1914.* London.

Polish Academy of Sciences (1961) *Problems of applied geography.* Institute of Geography, Warsaw.

Rakitnikov, A. N. (1970) *Geografiya sel'skogo khozyaystva.* Moscow.

Reeds, L. G. (ed.) (1973) *Agricultural typology and land use.* Proc. *Agricultural Typology Commission Meeting,* Hamilton.

Reining, C. C. (1966) *The Zande Scheme.* Evanston, Illinois.

Renner, G. T. (1935) 'The Statistical approach to regions'. *Annals Assoc. American Geographers,* Vol. 25, 137–152.

Rey, V. (1976) 'Visites dans les campagnes Chinoises'. *Annales de Geographie,* No. 470, 441–472.

Romney, D. H. (ed.) (1959) *Land in British Honduras.* H.M.S.O. London.

Rosen, G. (1975) *Peasant society in a changing economy; comparative development in Southeast Asia and India.* Urbana, Chicago, London.

Russell, E. W. (1973) *Soil conditions and plant growth.* London, 10th ed.

Samuelson, P. A. (1973) *Economics.* New York, 9th ed.

Sauer, C. O. (1919) 'Mapping the utilisation of the land'. *Geographical Review,* Vol. 8, 47–54.

Sauer, C. O. (1952) *Agricultural origins and dispersals.* American Geographical Society, New York.

Schlippe, Pierre de (1956) *Shifting cultivation in Africa: The Zande system of agriculture.* London.

Scott, P. (1957) 'The agricultural regions of Tasmania: a statistical definition'. *Economic Geography,* Vol. 33, 109–121.

Sears, P. D. (1962) 'Regional differences in grassland farming practice, New Zealand conditions and some overseas comparisons', in *Land and livelihood, geographical essays in honour of George Jobberns* (ed. M. McCaskill), 191–202. N.Z. Geog. Soc., Christchurch.

Selby, M. J. (1971) *The surface of the earth,* Vol. 2, *Climate, soils and vegetation.* London.

Siddle, D. J. (1970) 'Location theory and the subsistence economy; the spacing of rural settlements in Sierra Leone'. *Jnl. Trop. Geog.,* Vol. 31, 79–90.

Slicher van Bath, B. (1963) *The agrarian history of western Europe AD 500–1850.* London.

Smith, W. (1949) *An economic geography of Great Britain.* London.

Spencer, J. E., and Hale, G. A. (1961) The origin, nature and distribution of agricultural terracing, *Pacific Viewpoint,* Vol. 2, 1–40. Comment by A. C. S. Wright and reply by J. E. Spencer appear in *Pacific Viewpoint,* Vol. 3 (1962), 97–105.

Spencer, J. E., and Horvath, R. J. (1963) 'How does an agricultural region originate? *Annals Assoc. American Geographers,* Vol. 53, 74–92.

Spencer, J. E., and Stewart, N. R. (1973) 'The nature of agricultural systems'. *Annals Assoc. American Geographers,* Vol. 63, 529–544.

Stamp, L. D. (ed.) (1937–44) *The land of Britain,* The report of the Land Utilisation Survey of Britain.

Stamp, L. D. (1948) *The land of Britain, its use and misuse.* London, 3rd ed. 1962.

Stephens, N. (1963) 'Climate', in *Land use in Northern Ireland* (ed. L. Symons). London, 75–92.

Stephens, P. R. (1976) 'Farming', in *New Zealand atlas* (ed. I. Ward). Wellington, 144–150.

Stewart, G. A. (ed.) (1968) *Land evaluation.* Melbourne.

Stewart, O. C. (1956) 'Fire as the first great force employed by man', in *Man's role in changing the face of the earth* (ed. W. L. Thomas). Chicago, 115–133.

Storie, R. E. (1933) 'An index for rating the agricultural value of soils'. *Univ. Calif. Agric. Expt. Sta. Bull.*, 556.

Storie, R. E. (1948) 'Revision of soil rating chart'. *Univ. Calif. Agric. Expt. Sta. Bull.*

Stover, S. L. (1969) 'The government as farmer in New Zealand'. *Economic Geography*, Vol. 45, 324–338.

Sumner, B. H. (1947) *Survey of Russian history.* London, 2nd ed. 1961.

Symons, L. (ed.) (1963) *Land use in Northern Ireland.* London.

Symons, L. (1972) *Russian agriculture.* London.

Tagg, J. R. (1957) *Wind data related to the generation of electricity by wind power.* British Electrical and Allied Industries Research Association, Leatherhead.

Tarrant, J. R. (1969) 'Some spatial variations in Irish agriculture, *Tijdschrift voor Economische en Sociale Geografie*, Vol. 60, 228–237.

Tarrant, J. R. (1974) *Agricultural geography.* Newton Abbot.

Tax, S. (1956) 'Contribution to discussion on subsistence economies', in *Man's role in changing the face of the earth* (ed. W. L. Thomas). Chicago, 421.

Taylor, J. A. (ed.) (1970) *Weather economics.* Oxford.

Taylor, J. A. (ed.) (1974) *Climatic resources and economic activity.* Newton Abbot.

Taylor, N. H., and Pohlen, I. J. *See* New Zealand Department of Scientific and Industrial Research.

Tempany, A., and Grist, D. H. (1958) *An introduction to tropical agriculture.* London.

Thoman, R. S. (1962) *The geography of economic activity.* New York.

Thomas, D. (1963) *Agriculture in Wales during the Napoleonic wars.* Cardiff.

Thomas, W. L. (ed.) (1956) *Man's role in changing the face of the earth.* Chicago.

Thornthwaite, C. W. (1948) 'An approach toward a rational classification of climate'. *Geographical Review*, Vol. 38, 55–94.

Thünen, J. von (1826) *Der isolierte Staat in Beziehung auf Landwirtschaft und Nationalökonomie.* Rostock.

Tracy, M. (1964) *Agriculture in western Europe: crisis and adaptation since 1880.* London.

Trow-Smith, R. (1957) *A history of British livestock husbandry to 1700.* London.

Trow-Smith, R. (1959) *A history of British livestock husbandry 1700–1900.* London.

Ucko, P. J., and Dimbleby, G. W. (eds) (1969) *The domestication and exploitation of plants and animals.* London.

United Nations Food and Agriculture Organisation (1957) 'Shifting cultivation'. *Unasylva*, Vol. 11, No. 1, 9–11, reprinted in *Trop. Agric. Trin.*, Vol. 34, 159–164.

U.S.A. Dept. of Agriculture (1951) *Soil survey manual.* Washington.

U.S.A. Dept. of Agriculture (1954) *A manual on conservation of soil and water.* Washington.

U.S.A. Dept. of Agriculture (1955) *Water.* Washington.

U.S.A. Dept. of Agriculture (1957) *Soil.* Washington.

U.S.A. Dept. of Agriculture (1975) *Soil taxonomy: a basic system of soil classification for making and interpreting soil surveys.* Agric. Handbook No. 436, Washington.

U.S.A. National Resources Planning Board (1941) *Land classification in the United States.* Washington.

U.S.S.R. Academy of Sciences of the (1962) *Soil-geographical zoning of the U.S.S.R. (in relation to the agricultural usage of lands)* translated by A. Gourevitch. Jerusalem and London, 1963.

Vavilov, N. I. (ed. and principal author) (1935) *Teoreticheski osnovi selektsi rasteni.* Moscow, 1935. Selected writings translated in 'The origin variation, immunity and breeding of cultivated plants'. *Chronica Botanica,* Vol. 13, 1949–50, Waltham, Mass., 1951, 1–366.

Vilenskii, D. G. (1957) *Soil science.* Moscow, trans. by A. Barron and Z. S. Cole, Jerusalem and London, 1963.

von Thünen. *See* Thünen.

Voon Phin Keong (1972) 'Size aspects of rubber smallholdings in West Malaysia: a case study of Bentong, Pahang'. *Jnl. Trop. Geog.,* Vol. 34 65–76.

Wadham, S., Wilson, R., and Wood, J. (1957) *Land utilisation in Australia* Melbourne and London, 3rd ed.

Wareing, P. F., and Cooper, J. P. (eds) (1971) *Potential crop production.* London.

Watson, J. A. S., and More, J. A. (1962) *Agriculture.* Edinburgh and London, 11th ed.

Watters, R. F. (1960) 'The nature of shifting cultivation: a review of recent research'. *Pacific Viewpoint,* Vol. 1, 59–99.

Weaver, J. C. (1954) 'Crop combination regions in the Middle West'. *Geographical Review,* Vol. 44, 175–200.

Weber, A. (1909) *Theory of the location of industries,* translated by C. J. Freidrich, Chicago, 1929. 2nd impression, 1957.

Webster, C. C., and Wilson, P. N. (1966) *Agriculture in the tropics.* London

Weiers, C. J., and Reid, I. G. (1974) *Soil classification, land valuation, and taxation, 'the German experience'.* Wye College, Ashford.

White, C. (1975) 'The impact of Russian railway construction on the market for grain in the 1860s and 1870s', in *Russian transport, ar historical and geographical survey* (eds L. Symons and C. White). London, 1–45.

Whittlesey, D. (1936) 'Major agricultural regions of the earth'. *Annal. Assoc. American Geographers,* Vol. 26, 199–240.

Whittlesey, D. (1954) 'The regional concept and the regional method', in

American geography, inventory and prospect (eds P. E. James, C. F. Jones, and J. K. Wright). Syracuse.

Winter, E. H. (1956) *Bwamba economy.* East African Inst. of Social Research.

Wolpert, J. (1964) 'The decision process in spatial context'. *Annals Assoc. American Geographers,* Vol. 54, 537–558.

Wycherley, P. R. (1963) 'Variation in the performance of *Hevea* in Malaya'. *Jnl. Tropical Geography,* Vol. 17, 143–171.

Youngblood and Cox (1922) *An economic study of a typical ranching area on the Edwards plateau of Texas,* Bulletin No. 295, Texas Agricultural Experimental Station.

Zobler, L. (1957) 'Statistical testing of regional boundaries'. *Annals Assoc. American Geographers,* Vol. 47, 83–95.

Zuckerman, Sir Solly and others (1958) *Land ownership and resources.* Cambridge.

Zvorykin, K. V. (1963) 'Scientific principles for an agro-production classification of lands'. *Geografiya i khozyaystvo,* trans. in *Soviet Geography, Review and Translation,* Vol. 4, 3–10.

Index

Aberdeen-Angus cattle, 15, 107, 132
Ableiter, J. K., 248
aboriginals in Malaysia, 185–6
aerial spraying, *see* aviation
Africa, 8, 9, 19, 31, 59, 67, 69, 70, 84, 98, 101, 167, 252, *see also* individual countries
agrarian revolution, 11, 14 ff, 60, 151 ff, 225
agricultural atlases, of England and Wales, 204–5, 215
Agricultural Development Advisory Service (Lands), 117, 238
agricultural geography, ix, 1–2, 253–8
Agricultural Land Service, 237–8
Agricultural Marketing Acts, 109
agricultural regions, *see* regions
'agriculturalisation', 60, 87
agriculture: definition of, 1–2; history of, 7 ff, 60, 151–8, 225; origins of 6–10; plantation, *see* plantation agriculture
Agriculture Act, 1947, 110
aid, American food, 90
aircraft in agriculture, 19, 51, 54, 86, 128
air photography, in land use survey, 224, 233
air transport, 71
Alexis, Tsar, 151
alfalfa (lucerne), 11, 23, 105, 192, 236, 248
alienation of land, 182
allotments, horticultural, 226, 230
alluvial soils, 42, 182, 242
Alps, European, 27, 31, 47, 51
Alps, Southern, 121, 131, 133
altitude, agricultural effects of, 29, 46 ff, 131, 137, 241; *see also* hill land, mountains, etc
amalgamation of farm units, 17, 65, 75, 77, 111, 112, 119, 134, 148; *see also* consolidation of farms
Amazon region, 137, 173
America, 19, 73, 77, 97, 103, 122, 136; North, 16, 38, 90, 99, 125, 136, 138, 202 ff, 259; South and Central 9, 17, 59, 60, 70, 138; *see also* individual countries
Anderson, J., 4
Andean region, 9, 70
Anglo-Polish Geographical Seminar, 231
animals: domestication of, 8-10; herd, 8, 9; in the soil, 43–4; nutrition, on hills 50–1; *see also* livestock
Anuchin, V. A., 190
apples, 111, 119
Arab action on oil prices, 254
Arab traders in S. E. Asia, 135
arable farming: concentric zones, 194–6; erosion, 22, 235–6; historical, 8 ff; in British Isles, 102, 107 ff; in land classification, 235 ff; in land use survey, 226 ff; in New Zealand, 123; intensity of, 77; suitability of land for, 235 ff; types of, 94–5, 102; use for market gardening, 77–8; with communal grazings, 68; *see also* cereals, individual crops
Archimedes screw, 11
ard (light plough), 10
Argentine, 23, 99, 110, 122, 192
arid regions, 19, 23–4, 25–6, 30–31, 158, 221, 236
Armagh, Co., 119
Armstrong, W. J., 251–2
Ashanti, 173
Asia, 9, 19, 38, 60, 63, 67, 70, 98, 125, 167; central, 8, 9; south east, 8, 42, 56, 98, 135 ff; 163–4; Soviet, 103, 150, 164, 168; *see also* individual countries
asparagus, 22
Assam, 9
Auckland, North, 127
Australia, 16, 91, 121–2, 125, 132; arid areas, 3, 24, 26; exports, 70, 71, 73, 79; floods, 26; labour shortage, 82; land clearing, 86; types of farming, 97, 99

270